AS CHEMISTRY for AQA

Revision Guide

Sandra Clinton
Emma Poole

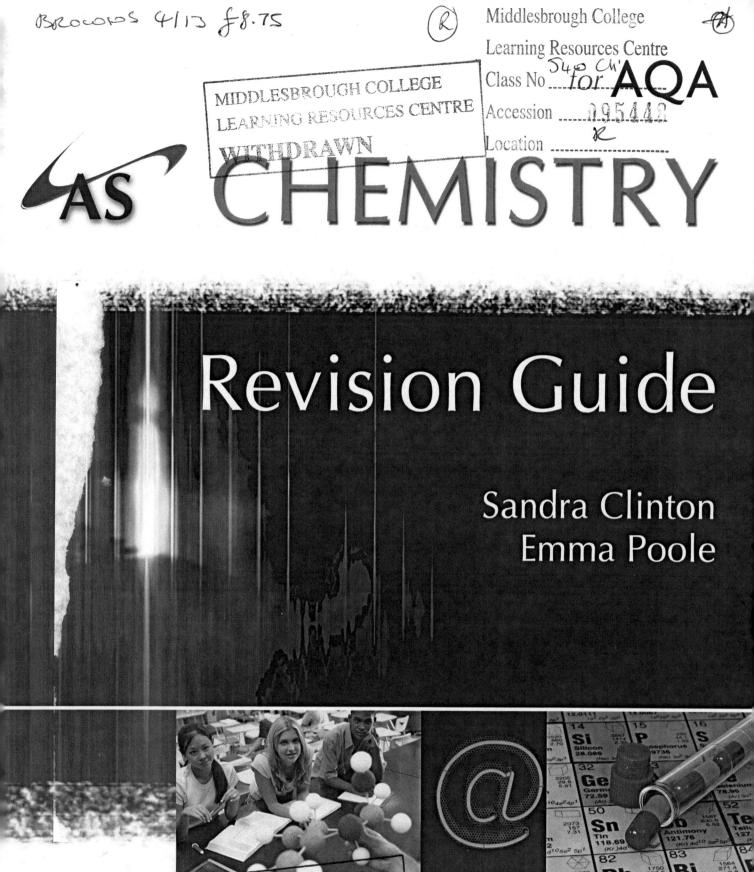

OXFORD
UNIVERSITY PRESS

OXFORD
UNIVERSITY PRESS

Great Clarendon Street, Oxford OX2 6DP

Oxford University Press is a department of the University of Oxford.
It furthers the University's objective of excellence in research, scholarship,
and education by publishing worldwide in

Oxford   New York

Auckland   Cape Town   Dar es Salaam   Hong Kong   Karachi
Kuala Lumpur   Madrid   Melbourne   Mexico City   Nairobi
New Delhi   Shanghai   Taipei   Toronto

With offices in

Argentina   Austria   Brazil   Chile   Czech Republic   France   Greece
Guatemala   Hungary   Italy   Japan   Poland   Portugal   Singapore
South Korea   Switzerland   Thailand   Turkey   Ukraine   Vietnam

© Oxford University Press 2008

The moral rights of the authors have been asserted

Database right Oxford University Press (maker)

First published 2008

British Library Cataloguing in Publication Data

Data available

ISBN: 978-0-19-915274-2

10 9 8 7 6 5 4

Printed in Great Britain by Bell and Bain Ltd, Glasgow

Paper used in the production of this book is a natural, recyclable product
made from wood grown in sustainable forests. The manufacturing process
conforms to the environmental regulations of the country of origin.

MIX
Paper from
responsible sources
FSC® C007785

Contents

		Page	Specification reference

WELCOME/INTRODUCTION

Welcome to the Revision Guide for AQA AS Chemistry.

We've tried to package the course in such a way as to help you more easily go into the examination room with added confidence for success.

You'll find details on:

- how AQA assess you through the examinations together with how examinations work
- there's all the content from the course presented in Specification order (with a handy list of Specification references for easy location)
- revision guidance together with Unit 'rough guides' to help you with your planning

There are practice exam-style questions together with quick-check style questions on each spread.

Sandra Clinton
Emma Poole
2008

ASSESSMENT IN AQA AS CHEMISTRY

Here are the nuts and bolts of the course: the practicalities of how AQA assesses the progress you have made during your studies.

To make the most of your exams it's essential you know how the various sections of the exams are constructed. If you know *how* the exams work you will be better placed to gain maximum marks.

Assessment overall

Your AS Chemistry qualification is made up from three Units. Units 1 and 2 are assessed by written examination; Unit 3 is an assessment of the investigative and practical skills you have gained during the course.

Unit	Name	Length of exam	% age AS marks	% age (total) A level marks
1	*Foundation chemistry*	1 hour 15 minutes	33.33%	16.66%
	4–6 short answer questions plus 1–2 longer structured questions			
2	*Chemistry in action*	1 hour 45 minutes	46.67%	23.33%
	6–8 short answer questions plus 2 longer structured questions			
3	*Investigative and practical skills in chemistry* (internal assessment)		20%	10%

Each of these three Units will examine your ability to meet the assessment objectives set out below. Work through the statements and highlight the key words.

- note that *these are skills*, not lists of content (which are found in the Specification)

Assessment objectives plus...

Assessment objectives AO1 and AO2 are assessed across all Units while AO3 is assessed mainly in Unit 3. In essence the assessment objectives AO1 and AO2 require you to know lots of stuff – facts, figures, and chemistry; whilst AO3 expects you to know *how* chemistry works – or what makes it tick – it's all about being a practical scientist, a chemist!

Assessment objective AO1

Knowledge and understanding of science and of *How Science Works*

You should be able to:

- recognise, recall, and show understanding of scientific knowledge
- select, organise, and communicate relevant information in a variety of forms

Assessment objective AO2

Application of knowledge and understanding of science and of *How Science Works*

You should be able to:

- analyse and evaluate scientific knowledge and processes
- apply scientific knowledge and processes to unfamiliar situations, including those related to issues
- assess the validity, reliability, and credibility of scientific information

Assessment objective AO3

You should be able to:

- demonstrate and describe ethical, safe and skilful practical techniques and processes, selecting appropriate qualitative and quantitative methods
- make, record and communicate reliable and valid observations and measurements with appropriate precision and accuracy
- analyse, interpret, explain and evaluate the methodology, results and impact of their own and others' experimental and investigative activities in a variety of ways

Quality of written communication

It's all very well knowing loads of stuff but you need to be able to communicate it and get your ideas across to the examiner. You might not think it but the examiner has your interests at the centre of their job – you need to give them the easiest route to maximizing your marks.

You should:

- ensure that text is legible and that spelling, punctuation, and grammar are accurate so that meaning is clear
- select and use a form and style of writing appropriate to purpose and to complex subject matter
- organise information clearly and coherently, using specialist vocabulary where appropriate

Quality of written communication is assessed across all externally assessed Units; if you write clear, well explained answers then you should obtain any marks assigned to it.

Investigative and practical skills

Your school will decide if your work is to be internally or externally marked – this has no impact on the work you will complete but does make the system appear slightly more complicated.

There are two routes: Route T and Route X.

***Route T*: marked inTernally by your Teachers.** You will complete two components:

- Practical **S**kills **A**ssessment (PSA)
- Investigative **S**kills **A**ssignment (ISA)

***Route X*: marked eXternally by the eXam board:** You will complete two components:

- Practical **S**kills **V**erification (PSV)
- **E**xternally **M**arked **P**ractical **A**ssignment (EMPA)

You will be prepared for these assessments during the course of your normal lessons. They will examine your ability to meet assessment objective AO3.

HOW EXAMINATIONS WORK

Part of your course involves understanding/knowing *how science works*. Well, for maximum marks in your exam you need to know *how examinations work*. And in much the same way as science there are in-built rules that form the foundation of how examinations are constructed.

Get and speak the *lingo*: know exam-speak, play the game. Here are some popular terms which are often used in exam questions. Make sure you know what each of these terms means. For a term that requires a written answer it is most unlikely that one/two words will do!

- **Calculate**: means calculate and write down the numerical answer to the question. Remember to include your working and the units.
- **Define**: write down what a chemical/term means. Remember to include any conditions involved.
- **Describe**: write down using words and, where appropriate diagrams, all the key points. Think about the number of marks that are available when you write your answer.
- **Discuss**: write down details of the points in the given topic.
- **Explain**: write down a supporting argument using your chemical knowledge. Think about the number of marks that are available when you write your answer.
- **List**: write down a number of points. Think about the number of points required. Remember, incorrect answers will cost you marks.
- **Sketch**: when this term is used a simple freehand answer is acceptable. Remember to make sure that you include any important labels.
- **State**: write down the answer. Remember a short answer rather than a long explanation is required.
- **Suggest**: use your chemical knowledge to answer the question. This term is used when there is more than one possible answer or when the question involves an unfamiliar context.

GETTING DOWN TO REVISION

OK, you're committed to preparing for your examination! How do *you* go about it? Remember, there are almost as many ways to revise as there are students revising. Underneath the methods of revising there are some common goals that the revising has to achieve.

1 Boost your confidence

Careful revision will enable you to perform at your best in your examinations. So give yourself the easiest route through the work. Organise the work into small, manageable chunks and set it out in a timetable. Then each time you finish a chunk you can say to yourself 'done it', and then move on to the next one. **And give yourself a reward too!** It's amazing how much of a lift it gives you by working in this way.

2 Be successful

To be successful in AS level chemistry you must be able to:
- recall information
- apply your knowledge to new and unfamiliar situations
- carry out precise and accurate experimental work*
- interpret and analyse both your own experimental data and that of others*

*experience gained with practical work will help you with answering questions in the examination so don't set aside all this valuable knowledge and understanding

How this revision guide can help

This book will provide you with the facts which you need to recall and some examination practice.

Use it as a working book and colour it in! Start with the Unit guides/maps that show you each of the topics covered. Read them and highlight the areas which you are already confident about. Then, in a different colour mark off the sections one at a time as your revision progresses. The more colour on your map the more work you have done to prepare for your examination. By doing this you will feel positive about what you have achieved rather than negative about what you still have to do.

For your revision programme you might like to use some or all of the following strategies:
- read through the topics one at a time and try the quick questions
- choose a topic and make your own condensed summary notes
- colour important diagrams
- highlight key definitions or write them onto flash cards

- work through facts until you can recall them
- constantly test your recall by covering up sections and writing them from memory
- ask your friends and family to test your recall
- make posters for your bedroom walls
- use the 'objectives' as a self-test list
- carry out exam practice
- work carefully through the material on each page, highlighting the guide/map sections as you go
- make 'to do' lists like those on the exam practice pages
- and don't forget about HSW (How Science Works)!

Whatever strategies you use, measure your revision in terms of the progress you are making rather than the length of time you have spent working. You will feel much more positive if you are able to say specific things you have achieved at the end of a day's revision rather than thinking, 'I spent eight hours inside on a sunny day!'. Don't sit for extended periods of time. Plan your day so that you have regular breaks, fresh air, and things to look forward too.

Watch out: revision is an active occupation - just reading information is not enough! You will need to be active in your work for your revision to be successful.

Improving your recall: A good strategy for recalling information is to focus on a small number of facts for five minutes. Copy out the facts repeatedly in silence for five minutes then turn your piece of paper over and write them from memory. If you get any wrong then just write these out for five minutes. Finally test your recall of all the facts. Come back to the same facts later in the day and test your self again. Then revisit them the next day and again later in the week. By carrying out this process they will become part of your long term memory – you will have learnt them!

Past paper practice: Once you have built up a solid factual knowledge base you need to test it by completing past paper practice. It might be a good idea to tackle several questions on the same topic from a number of papers rather than working through a whole paper at once. This will enable you to identify any weak areas so that you can work on them in more detail. Finally, remember to complete some mock exam papers under exam conditions.

A final word (or two) for the examination room

Unlike the GCSE examination this examination does not have a foundation tier and a higher tier so you must be prepared to answer questions on all the topics outlined in the Specification. Here are some obvious, and not so obvious thoughts:

- read through all the questions (*obvious*)
- identify which questions you can answer well (*obvious*)
  o start by answering these questions (*not so obvious*)
- once you have read a question carefully make sure you answer that question and NOT something you might think is the question (*not so obvious*)
- look at the number of marks that are available for each question and take this into account when you write your answer (*obvious and not so obvious*)
- answer space: the amount of space left for the answer will give you an indication of the length of answer the examiner expects (*obvious*)
  o short space = short answer (*obvious*)
  o longer space(1) = extended answer probably required; perhaps a sentence, or two; a calculation with working; a list containing a selection items (*obvious*)
  o longer space(2) = one word answer unlikely to be sufficient (*not so obvious*)
- answer all the questions, even if you have to guess at some (*obvious*)
- pace yourself and try to leave enough time to check your answers at the end (*obvious*)

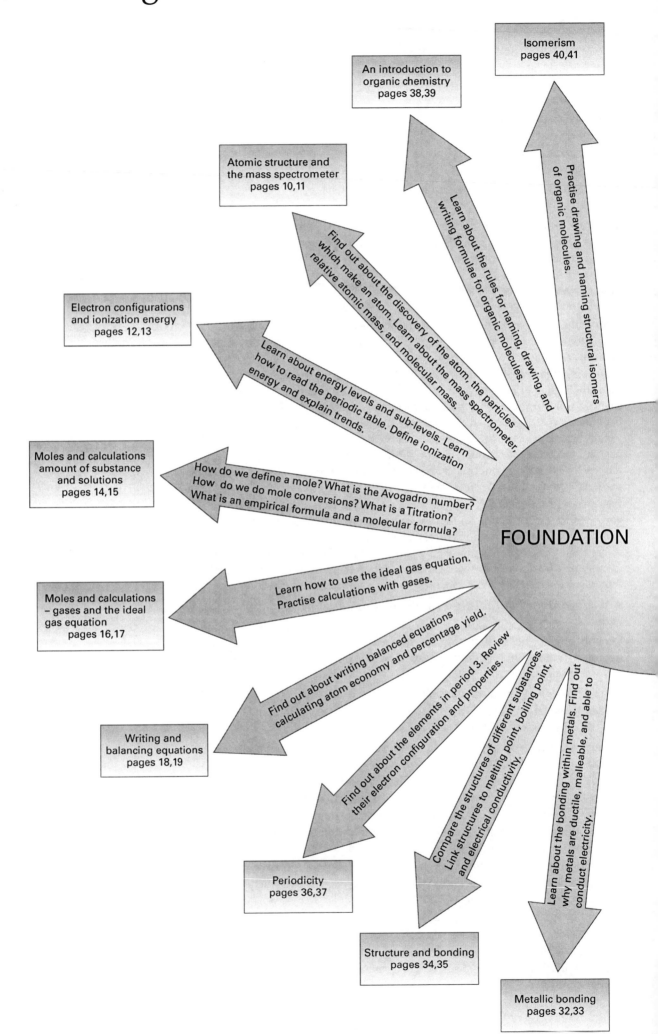

Isomerism
pages 40,41

An introduction to
organic chemistry
pages 38,39

Atomic structure and
the mass spectrometer
pages 10,11

Practise drawing and naming structural isomers
of organic molecules.

Learn about the rules for naming, drawing, and
writing formulae for organic molecules.

Find out about the discovery of the atom, the particles
which make an atom. Learn about the mass spectrometer,
relative atomic mass, and molecular mass.

Electron configurations
and ionization energy
pages 12,13

Learn about energy levels and sub-levels. Learn
how to read the periodic table. Define ionization
energy and explain trends.

Moles and calculations
amount of substance
and solutions
pages 14,15

How do we define a mole? What is the Avogadro number?
How do we do mole conversions? What is a Titration?
What is an empirical formula and a molecular formula?

FOUNDATION

Learn how to use the ideal gas equation.
Practise calculations with gases.

Moles and calculations
– gases and the ideal
gas equation
pages 16,17

Find out about writing balanced equations
calculating atom economy and percentage yield.

Writing and
balancing equations
pages 18,19

Find out about the elements in period 3. Review
their electron configuration and properties.

Compare the structures of different substances.
Link structures to melting point, boiling point,
and electrical conductivity.

Learn about the bonding within metals. Find out
why metals are ductile, malleable, and able to
conduct electricity.

Periodicity
pages 36,37

Structure and bonding
pages 34,35

Metallic bonding
pages 32,33

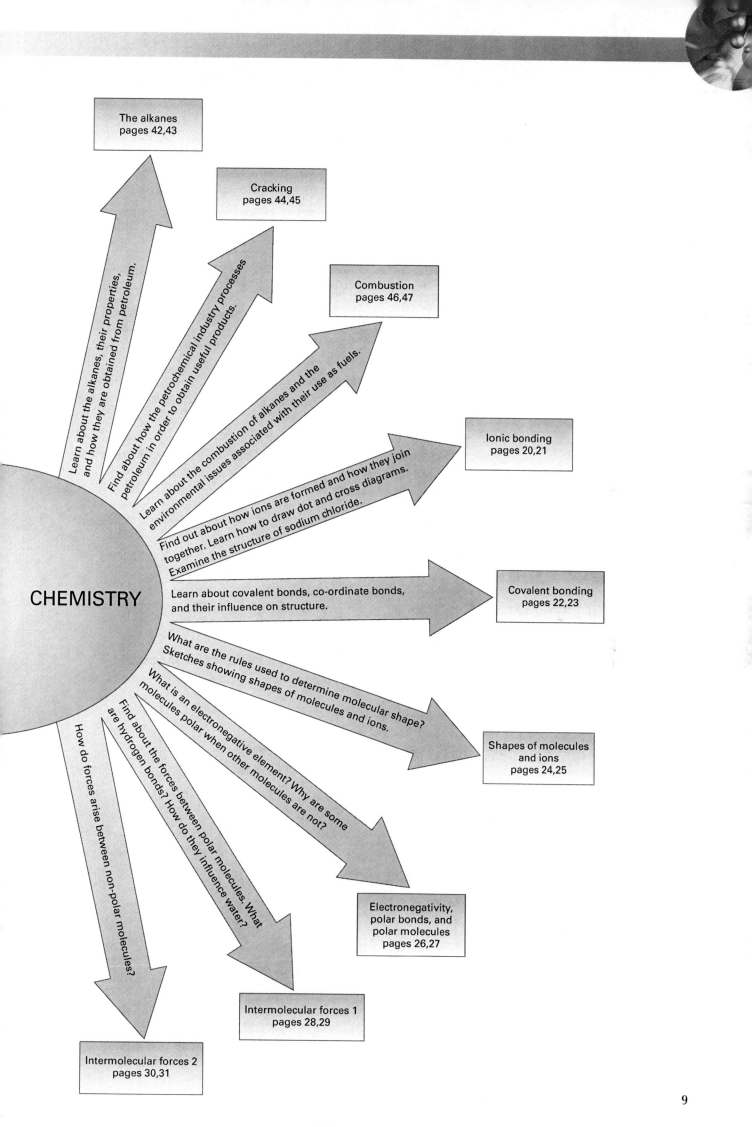

CHEMISTRY

The alkanes
pages 42,43

Cracking
pages 44,45

Combustion
pages 46,47

Ionic bonding
pages 20,21

Covalent bonding
pages 22,23

Shapes of molecules
and ions
pages 24,25

Electronegativity,
polar bonds, and
polar molecules
pages 26,27

Intermolecular forces 1
pages 28,29

Intermolecular forces 2
pages 30,31

Learn about the alkanes, their properties, and how they are obtained from petroleum.

Find about how the petrochemical industry processes petroleum in order to obtain useful products.

Learn about the combustion of alkanes and the environmental issues associated with their use as fuels.

Find out about how ions are formed and how they join together. Learn how to draw dot and cross diagrams. Examine the structure of sodium chloride.

Learn about covalent bonds, co-ordinate bonds, and their influence on structure.

What are the rules used to determine molecular shape? Sketches showing shapes of molecules and ions.

What is an electronegative element? Why are some molecules polar when other molecules are not?

Find about the forces between polar molecules. What are hydrogen bonds? How do they influence water?

How do forces arise between non-polar molecules?

1.01 Atomic structure and the mass spectrometer

OBJECTIVES

Atoms and sub-atomic particles

Atoms are made from sub-atomic or fundamental particles called **protons**, **neutrons**, and **electrons**. The masses and charges of these fundamental particles are very small so we look at them relative to the proton.

particle	relative mass	relative charge
proton	1	+1
neutron	1	0
electron	1/1836	−1

The mass of an electron is so small it is often considered to be negligible. Protons and neutrons are located in the nucleus of the atom while electrons orbit the nucleus.

Atomic number and mass number

Ensure that you can recall the definitions of atomic number and mass number.
- Atoms can be represented: $^A_Z X$
- Number of protons = atomic number (Z)
- Number of electrons = atomic number (Z) (ONLY for neutral atoms)
- Number of neutrons = mass number – atomic number ($A - Z$)

For ions (particles which have lost or gained electrons) the charge needs to be taken into account.

Beryllium atoms $^9_4 Be$ protons = 4, electrons = 4, neutrons = 9 − 4 = 5

Fluoride ions $^{19}_9 F^-$ protons = 9, electrons = 9 + 1 = 10, neutrons = 19 − 9 = 10

Calcium ions $^{40}_{20} Ca^{2+}$ protons = 20, electrons = 20 − 2 = 18, neutrons = 40 − 20 = 20

Isotopes

Make sure that you can remember the definition of an **isotope**.
- Isotopes contain differing numbers of neutrons but the same number of protons.
  - As a result they are chemically identical.

There are two isotopes of chlorine

$^{37}_{17} Cl$ protons = 17, electrons = 17, neutrons = 20

$^{35}_{17} Cl$ protons = 17, electrons = 17, neutrons = 18

By the end of this section you should be able to:

○ define the terms: atomic number, mass number, isotope

○ determine the number of protons, neutrons, and electrons present in an atom or ion

○ describe the steps in the mass spectroscopy experiment

○ interpret a simple mass spectrum

○ define relative atomic mass, relative molecular mass, and relative isotopic mass

○ calculate relative atomic mass

Atomic number, mass number and isotopes

Atomic number (Z): The number of protons in the nucleus.

Mass number (A): The number of protons and neutrons in the nucleus.

Isotope: Atoms that have the same number of protons but different numbers of neutrons in their nucleus.

HSW: Early models of the atom

The model of the atom has been developed over many hundreds of years. Scientists are constantly working to improve this model.

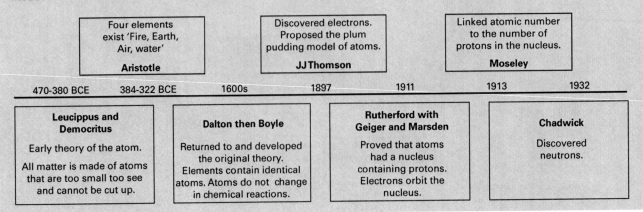

	Four elements exist 'Fire, Earth, Air, water'		Discovered electrons. Proposed the plum pudding model of atoms.		Linked atomic number to the number of protons in the nucleus.	
	Aristotle		**JJ Thomson**		**Moseley**	

470-380 BCE	384-322 BCE	1600s	1897	1911	1913	1932

Leucippus and Democritus	**Dalton then Boyle**	**Rutherford with Geiger and Marsden**	**Chadwick**
Early theory of the atom. All matter is made of atoms that are too small too see and cannot be cut up.	Returned to and developed the original theory. Elements contain identical atoms. Atoms do not change in chemical reactions.	Proved that atoms had a nucleus containing protons. Electrons orbit the nucleus.	Discovered neutrons.

The mass spectrometer

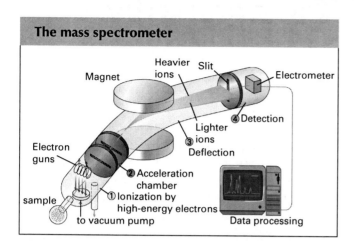

The mass spectrum of lead

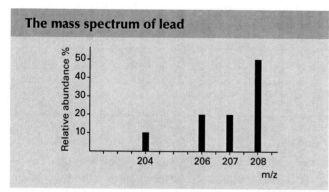

The mass spectrum of propanol

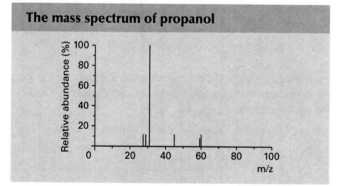

Definitions

Relative atomic mass

$$A_r = \frac{\text{mean mass of an atom of the element}}{\text{mass of one atom of }^{12}\text{C}} \times 12$$

Relative molecular mass

$$M_r = \frac{\text{mean mass of a molecule}}{\text{mass of one atom of }^{12}\text{C}} \times 12$$

Relative isotopic mass

$$= \frac{\text{mass of one atom of the isotope}}{\text{mass of one atom of }^{12}\text{C}} \times 12$$

Mass spectrometer

The mass spectrometer is a very sensitive machine used to analyse samples of elements in terms of the isotopes they contain and their relative amounts. The mass spectroscopy experiment is always carried out on gaseous samples under vacuum (so that no air is present). You must learn names and brief explanations for each step in the mass spectroscopy experiment.

- **Ionization** An electron gun is fired at the sample. This knocks off an outer shell electron forming a positive ion.
- **Acceleration** The positive ions are passed through an electric field (this also focuses the ions into a narrow beam).
- **Deflection** The fast moving positive ions are deflected by a strong magnetic field. Lighter ions are deflected more than heavier ions.
- **Detection** The positive ions are detected using an ion counter.

Relative atomic mass, A_r

The mass of all atoms is measured relative to the mass of a ^{12}C atom. Relative atomic mass can be calculated using the data from a mass spectrum. From the mass spectrum of lead opposite we obtain the following data:

m/z	204	206	207	208
relative abundance (%)	10	20	20	50

- m/z is the same as the mass of the ion if the charge is +1.
- relative abundance tells us the proportion of each isotope present in the sample.
- the A_r of lead can then be calculated as a weighted mean average.

$$A_r = \frac{(204 \times 10) + (206 \times 20) + (207 \times 20) + (208 \times 50)}{100} = 207$$

Determining relative molecular mass, M_r

Relative molecular mass is the mean mass of a molecule relative to the mass of a ^{12}C atom.

- If a molecule is placed in the mass spectrometer then the molecule will break up into smaller fragments or even its component elements.
- The complete molecule in the form of a molecular ion (the complete molecule minus an electron) will pass through the spectrometer.
- The peak for the molecular ion is the furthest peak to the right (i.e. the heaviest peak) on the mass spectrum.

For the mass spectrum of propanol the peak at $m/z = 60$ is for the molecular ion.

Questions

1 Write down the numbers of protons, neutrons and electrons in the following:

  a $^{23}_{11}\text{Na}$ b $^{16}_{8}\text{O}^{2-}$ c $^{80}_{35}\text{Br}$ d $^{40}_{19}\text{K}^+$

2 Write out the steps in the mass spectroscopy experiment five times. Test yourself.

3 Use the spectra data to determine the relative atomic mass and identity of the atom.

m/z	20	21	22
relative abundance (%)	90.9	0.3	8.8

11

1.02 Electron configurations and ionization energy

By the end of this section you should be able to:

○ sketch s and p orbitals

○ write electron configurations for atoms and ions up to krypton

○ define ionization energy

○ recall and interpret trends in ionization energy for period 3 and group 2

Shapes of orbitals

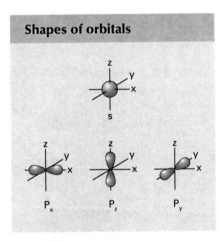

Configuration examples

- Potassium has 19 electrons. Its electron configuration is:
  $1s^2 2s^2 2p^6 3s^2 3p^6 4s^1$
- Scandium has 21 electrons. Its electron configuration is:
  $1s^2 2s^2 2p^6 3s^2 3p^6 4s^2 3d^1$
- A noble gas core is written in place of the inner shell electrons. This is shown using square brackets.
- Vanadium would be written $[Ar] 4s^2 3d^3$ since the electron configuration of argon is $1s^2 2s^2 2p^6 3s^2 3p^6$.
- Sodium could be written as $[Ne] 3s^1$.
- **Exception 1:** chromium has the configuration $[Ar] 3d^5 4s^1$ which is a more stable than $[Ar] 3d^4 4s^2$.
- **Exception 2:** copper has the configuration $[Ar] 3d^{10} 4s^1$ which is more stable than $[Ar] 3d^9 4s^2$.

Energy levels

Electrons are in constant motion around the nucleus of an atom.

- Electrons are found in **energy levels**.
- These are split into **sub-levels** with different maximum numbers of electrons.
- Different types of sub-level contain different numbers of **orbitals**.
- Each orbital can hold two electrons (one spinning up and one spinning down).
- Electrons are constantly moving so their exact location cannot be identified.
- The shape of an orbital tells you where an electron is most likely to be found.

energy level	1	2		3			4			
type of sub-level	s	s	p	s	p	d	s	p	d	f
number of orbitals in sub-level	1	1	3	1	3	5	1	3	5	7
maximum number of electrons in sub-level	2	2	6	2	6	10	2	6	10	14
maximum number of electrons in level	2	8		18			32			

Electron configurations

Use three principles for writing electron configurations. Work through each rule considering the examples shown in the table below. You won't be expected to name the rules in the examination but make sure you can apply each of them.

The 4s orbital is slightly lower in energy than the 3d so is filled first.

- The Aufbau principle – electrons fill energy levels in order of increasing energy.
- The Pauli exclusion principle – each orbital can contain a maximum of two electrons. Electrons in the same orbital must have opposite spins. This is shown using an up arrow and a down arrow.
- Hund's rule – orbitals are occupied unpaired before paired

Element	Electron configuration	Spin diagrams								
		1s	2s	2p			3s	3p		
H	$1s^1$	↑								
He	$1s^2$	↑↓								
Li	$1s^2 2s^1$	↑↓	↑							
C	$1s^2 2s^2 2p^2$	↑↓	↑↓	↑	↑					
O	$1s^2 2s^2 2p^4$	↑↓	↑↓	↑↓	↑	↑				
P	$1s^2 2s^2 2p^6 3s^2 3p^3$	↑↓	↑↓	↑↓	↑↓	↑↓	↑↓	↑	↑	↑

- The period that an element is in determines its highest energy level.
- The group that an element is in determines its outer electron configuration.

element	period	highest energy level	block and group	outer electron configuration	electron configuration
Na	3	3	s block group 1	s^1	$1s^2 2s^2 2p^6 3s^1$
N	2	2	p block group 5	$s^2 p^3$	$1s^2 2s^2 2p^3$
V	3	3	d block	d^3	$1s^2 2s^2 2p^6 3s^2 3p^6 4s^2 3d^3$

Electron configurations of ions

element	lithium	aluminium	oxygen	fluorine
group	1	3	6	7
electron configuration of atom	$1s^2\ 2s^1$	$1s^2\ 2s^2\ 2p^6$ $3s^2\ 3p^1$	$1s^2\ 2s^2$ $2p^4$	$1s^2\ 2s^2$ $2p^5$
electron configuration of ion	$1s^2$	$1s^2\ 2s^2\ 2p^6$	$1s^2\ 2s^2$ $2p^6$	$1s^2\ 2s^2$ $2p^6$
symbol of ion	Li^+	Al^{3+}	O^{2-}	F^-

Ions of d block elements

- Iron has the configuration $[Ar]\ 3d^6\ 4s^2$: its ions, Fe^{2+} and Fe^{3+} are $[Ar]\ 3d^6$ and $[Ar]\ 3d^5$.

Ionization energy

The values obtained for ionization energies provide substantial evidence for the existence of energy levels and sub-levels. Make sure you memorize the definition for ionization energy.

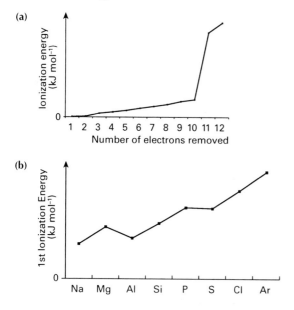

(a) Successive ionization energies of magnesium (b) First ionization energies of the period 3 elements

Ionization energy

First ionization energy is the energy required to remove 1 electron from each atom in 1 mole of gaseous atoms forming 1 mole of ions with a single positive charge.

e.g. The equation for the first ionization (of magnesium) is:

$$Mg(g) \rightarrow Mg^+(g) + e^-$$

First ionization energies of group 2 elements

Ionization energy is influenced by

- the number of protons in the nucleus of an atom
- the distance of the outermost electron from the nucleus (the further it is from the nucleus the more the outer electron is shielded from the effect of the nucleus)
  - ® As a result the first ionization energy decreases as you go down a group.

Successive ionization energies of magnesium

Successive ionization energies provide further evidence for the existence of energy levels.

- There is a general increase in the energy needed to remove each electron from magnesium.
- This is because the electron is being removed from an ion with an increasing positive charge.
- There is a very big increase between the tenth and eleventh ionization energies as the eleventh electron is being removed from a full orbital that is nearer to the nucleus and of lower energy than the tenth electron.
- The group to which an element belongs can be determined by identifying where the big jump in ionization energy occurs.

First ionization energies of the elements in period 3

The graph of first ionization energies for period 3 elements provides substantial support for the existence of energy sub-levels.

- The number of protons increases across period 3 so there is an increase in the charge on the nucleus.
  - ® As a result the force of attraction between the nucleus and the outer electron increases.
- The number of electrons also increases but these go into the same energy level so are at the same distance from the nucleus and experience the same shielding.
  - ® As a result there is an increase in first ionization energy across period 3.

Drops in ionization energy occur at two points on the graph.

- Between magnesium and aluminium there is a decrease in ionization energy because the outer electron in aluminium is in a p sub-level which is of slightly higher energy than the s sub-level.

  Mg $1s^2\ 2s^2\ 2p^6\ 3s^2$ Al $1s^2\ 2s^2\ 2p^6\ 3s^2\ 3p^1$

- Between phosphorus and sulfur there is a slight decrease because in sulfur two of the p electrons are paired and this pair repel each other.

  P $1s^2\ 2s^2\ 2p^6\ 3s^2\ 3p^3$ S $1s^2\ 2s^2\ 2p^6\ 3s^2\ 3p^4$

Questions

1. Draw out an electron spin diagram and then write electron configurations using a noble gas abbreviation for:
   a Na   b S   c Cl   d Be
2. Write full electron configurations for a:
   a magnesium ion   b sulfide ion
   c oxide ion   d sodium ion
3. Sketch a graph showing the successive ionization energies of oxygen. Label the graph in detail, explaining the patterns it shows.

OBJECTIVES

By the end of this section you should be able to:

○ *define the term mole*

○ *understand what is meant by the Avogadro constant*

○ *calculate moles of solids and solutions*

○ *calculate empirical formulae*

○ *understand and use the term concentration*

Moles and the Avogadro constant

One mole is the amount of a substance that contains the same number of particles as there are atoms in exactly 12 g of ^{12}C.

The Avogadro constant (6.022×10^{23}) is the number of particles in one mole of a substance.

HSW: *Amedeo Avogadro was an Italian scientist whose work was not accepted until after his death.*

Amount of substance

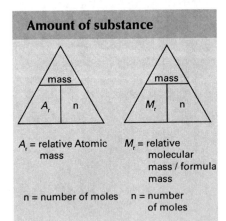

A_r = relative Atomic mass

M_r = relative molecular mass / formula mass

n = number of moles

n = number of moles

Empirical formulae and molecular formulae

The empirical formula shows the simplest whole number ratio of the atoms of each element in a compound.

The molecular formula shows the actual number of atoms of each element in one molecule of the compound. The molecular formula is a whole number multiple of the empirical formula.

The mole and the Avogadro constant

Chemists are interested in how many atoms, molecules, ions, etc. take part in reactions. All of these particles are very small so we cannot determine their mass. Instead we carry out calculations using the concept of the **mole**.

- The mole is the unit for amount of substance.
- One mole of a substance contains the same number of particles as there are atoms in exactly 12 g of ^{12}C.
- The number of particles in one mole of a substance is the **Avogadro constant** (after Amedeo Avogadro), 6.022×10^{23}.

Amount of substance

The amount of a substance is readily calculated from the relative atomic mass or the relative molecular mass of a substance using the relationship:

mass = $M_r \times n$

You must be able to manipulate this relationship in order to calculate mass, M_r, or n given suitable data.

Calculations using mass = $M_r \times n$

a Calculate the mass in grams of 2 moles of Mg atoms.

$$\text{mass} = A_r \times n = 24.3 \times 2 = 48.6 \text{ g}$$

b Calculate the mass in grams of 0.74 moles of $Ca(OH)_2$.

$$\text{mass} = M_r \times n = [40.1 + 2(17)] \times 0.74 = 74.1 \times 0.74 = 54.8 \text{ g}$$

c Calculate the number of moles of Na atoms in 6.5 g of Na atoms.

$$n = \text{mass}/A_r = 6.5/23.1 = 0.28$$

d Calculate the number of moles of MgO in 2×10^{-6} g of MgO.

$$n = \text{mass}/M_r = 2 \times 10^{-6}/40.3 = 4.96 \times 10^{-8}$$

e Calculate the M_r of aluminium oxide, Al_2O_3, if 0.5 moles of aluminium oxide has a mass of 51 g

$$M_r = \text{mass}/n = 51/0.5 = 102$$

Empirical and molecular formulae

The **empirical formula** of a compound is the simplest whole number ratio of the atoms of the elements in a compound. The **molecular formula** is the actual number of atoms.

For example: The empirical formula of butene is CH_2

The molecular formula of butene is C_4H_8

The empirical formula of a compound can be calculated if the percentage composition of the compound is known.

Calculating empirical formula

A compound contains 40.06% calcium, 11.99% carbon, and 47.95% oxygen by mass. Determine its empirical formula.

It is essential to show all your working in these calculations.

element	Ca	C	O
% by mass	40.06	11.99	47.95
÷ A_r	0.999	0.999	2.997
÷ smallest	1	1	3

Empirical formula: $CaCO_3$

Concentrations

A great deal of chemistry is carried out in aqueous solution. So a good understanding of the term concentration is essential for success at AS level.

- **Concentration** is the amount of a substance in moles dissolved in 1 dm³ of water (1 dm³ = 1000 cm³).
- This means that a 2 mol dm⁻³ solution of sulfuric acid contains 2 moles of sulfuric acid dissolved in 1 dm³ of water.
- This is 2 × 98.1 = 196.2 g of sulfuric acid.

The relationship between concentration and number of moles is: $n = c \times v$

n = number of moles, c = concentration in dm³, v = volume of solution in dm³

Note – The volume used in the concentration expression is needed in dm³. To convert dm³ to cm³ you must multiply by 1000. To convert cm³ to dm³ you must divide by 1000.

Acid–base titrations

Chemists use **titrations** to find the concentration of a reactant in solution. At As level you only need to know about acid-base titrations.

- A solution of known concentration (a standard solution) is reacted with the solution of unknown concentration.
- Very precise apparatus (burettes and pipettes) is used to measure the volumes of solutions.
- An indication is used to enable the point of exact neutralization to be determined.

In your examination you will only be expected to carry out calculations with monoprotic acids and bases (these donate/accept only one H⁺ ion).

Acid–base titration calculations

Follow these systematic steps to carry out a titration calculation. In an examination question you are likely to be guided through some of these steps. Make sure you practise lots of these!

25.0 cm³ of sodium hydroxide solution is exactly neutralized by 21.40 cm³ of 0.5 mol dm⁻³ hydrochloric acid. What was the concentration of the sodium hydroxide solution?

Step 1	Write a balanced equation NaOH(aq) + HCl(aq) → NaCl(aq) + H_2O(l)
Step 2	Write out the data given in the question under the equation. NaOH(aq) + HCl(aq) → NaCl(aq) + H_2O(l) 25.0 cm³ 21.40 cm³ ? 0.5 mol dm⁻³
Step 3	Convert the substance you know the most about into moles. number of moles of HCl(aq): $n = c \times v = 0.5 \times (21.4 \div 1000) = 0.0107$
Step 4	Use the balanced equation to determine the moles of the unknown substance. 1 mol HCl reacts with 1 mol NaOH. So number of moles of NaOH = 0.0107
Step 5	Convert the number into the units asked for in the question. concentration of sodium hydroxide solution: $c = n/v = 0.0107/(25.0 \div 1000) = 0.428$ mol dm⁻³

Solutions

Concentration is the amount of a substance in moles dissolved in 1 dm³ of water.

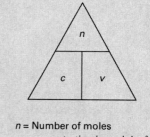

n = Number of moles
c = concentration in mol dm⁻³
v = volume of solution in dm³

Questions

1 Calculate the following. Remember to show all your working.

   a the number of moles of CaO in 4 g of CaO

   b the mass of 0.15 moles of K

   c the mass of 0.32 moles of LiOH

   d the concentration of a solution made by dissolving 6 g of NaOH in 1 dm³ of water

   e the concentration of a solution made by dissolving 3.2 g of HCl in 300 cm³ of water (note you need to use two of the mole relationships here)

   f the number of moles of H_2SO_4 present in 25.0 cm³ of 0.1 mol dm⁻³ H_2SO_4(aq)

   g the number of moles of HNO_3 present in 21.7 cm³ of 0.2 mol dm⁻³ HNO_3(aq)

   h the volume of 2 mol dm⁻³ KOH needed to provide 0.25 mol of KOH

2 25.0 cm³ of NaOH(aq) is exactly neutralized by 27.60 cm³ of 0.15 mol dm⁻³ HNO_3(aq). Calculate the concentration of the NaOH(aq) solution.

HSW: Absolute zero

This is the temperature at which a gas has zero volume.

This is at −273°C or O K.

Converting quantities

Temperatures

0°C = 273 K

To convert from °C to K add 273

To convert from K to °C subtract 273

Volumes

$1\ cm^3 = 10^{-6}\ m^3$

To convert from cm^3 to m^3 multiply by 10^{-6}

To convert from m^3 to cm^3 divide by 10^{-6}

Pressures

$1\ kPa = 10^3\ Pa$

To convert from kPa to Pa multiply by 10^3

To convert from Pa to kPa divide by 10^3

Standard conditions

Standard temperature is 273 K

Standard pressure is 1×10^5 Pa

These conditions are stated as STP.

Combining volumes of gases

In 1809 a French chemist Joseph Gay-Lussac proposed the law of combining gas volumes. This law states that, at constant pressure and temperature, the volumes of reacting gases and their gaseous products can be simplified as ratios of small whole numbers.

Using this law makes calculations involving gases very straightforward at constant temperatures and pressures.

Calculations with volumes of gases

What volume of hydrogen gas will be formed when 20 cm³ of methane reacts completely with steam at constant temperature and pressure?

$$CH_4(g) + H_2O(g) \rightarrow CO(g) + 3H_2(g)$$

Ratio of CH_4 to H_2 is 1:3

Volume of hydrogen formed = $20 \times 3 = 60\ cm^3$

The ideal gas equation

The phrase **molar volume**, V_m, is the volume occupied by one mole of gas.

- Molar volume increases as temperature increases.
- Molar volume decreases as pressure increases.
- The ideal gas equation links the volume of a gas to moles, pressure, and temperature of the gas.

The **ideal gas equation** is always stated in terms of pressure. You must be able to use it to calculate pressure, volume, and temperature. Make sure that you can write it in the four arrangements shown in the box.

Four arrangements of the ideal gas equation

$pV = nRT$

$p = \dfrac{nRT}{V}$

$V = \dfrac{nRT}{p}$

$T = \dfrac{pV}{nR}$

p : pressure in pascals (Pa)

V: volume in m^3

n: amount of a gas in moles

R: gas constant = 8.31 J K^{-1} mol^{-1}

T: temperature in K

Finding the pressure of a gas

What is the pressure exerted by 0.2 moles of chlorine gas in a vessel of 6 m³ at a temperature of 298 K?

$$p = \frac{nRT}{V} = \frac{0.2 \times 8.31 \times 298}{6} = 82.5\ Pa$$

Finding the volume of a gas

What is the volume of 6.4×10^{-4} moles of hydrogen gas at a pressure of 1 Pa at 273 K?

$$V = \frac{nRT}{p} = \frac{6.4 \times 10^{-4} \times 8.31 \times 273}{1} = 1.45\ m^3$$

Finding the temperature of a gas

At what temperature will 0.1 moles of oxygen gas occupy a volume of 24.4 m³ at a pressure of 2 Pa?

$$T = \frac{pV}{nR} = \frac{2 \times 24.4}{0.1 \times 8.31} = 58.7 \text{ K}$$

The ideal gas equation can be used in combination with the other mole relationships. You need to be able to decide which relationship to use. It is important that you set out your method clearly showing all the steps.

Calculating the M_r of a gas

At 25°C and 1×10^5 Pa pressure, 0.25 g of a gas occupies a volume of 87.2 cm³. What is the M_r of the gas?

Step 1 Ensure all the quantities are in the correct units.

$$\text{Volume: } 87.2 \text{ cm}^3 = 87.2 \times 10^{-6} \text{ m}^3$$

$$\text{Temperature: } 25°C = 25 + 273 = 298 \text{ K}$$

Step 2 Write out the re-arranged ideal gas equation.

$$n = \frac{pV}{RT}$$

$$n = \frac{(1 \times 10^5) \times (87.2 \times 10^{-6})}{8.31 \times 298} = 3.52 \times 10^{-3}$$

Step 3 Convert the number of moles into M_r using $M_r = mass/n$

$$M_r = 0.25 / (3.52 \times 10^{-3}) = 71$$

Standard conditions

Since the molar volume of a gas is dependent on temperature and pressure it is normally measured at **standard temperature and pressure, STP**.

- standard temperature is 273 K
- standard pressure is 1×10^5 Pa

Using the ideal gas equation the molar volume at STP can be calculated

$$V = \frac{nRT}{p} = \frac{1 \times 8.31 \times 273}{1 \times 10^5} = 0.0227 \text{ m}^3 = 22.7 \text{ dm}^3$$

Therefore one mole of any gas will occupy 22.7 dm³ at a temperature of 273 K and a pressure of 1×10^5 Pa.

Calculating the volume of gas released in a reaction

2 g of calcium carbonate is reacted with an excess of 0.2 mol dm⁻³ hydrochloric acid. Calculate the volume of carbon dioxide (in m³) released at standard temperature and pressure. $CaCO_3(s) + 2HCl(aq) \rightarrow CaCl_2(aq) + H_2O(l) + CO_2(g)$

Step 1 Calculate the number of moles of calcium carbonate.

$$n = mass/M_r = 2/(40.1 + 12 + (3 \times 16)) = 2/100.1 = 0.01998$$

Step 2 Use the balanced equation to determine the number of moles of carbon dioxide.

$$\text{ratio } CaCO_3 : CO_2 \qquad 1:1$$

$$n \, CO_2 = 0.01998$$

Step 3 Use the ideal gas equation to calculate the volume of carbon dioxide.

$$V = \frac{nRT}{p} = \frac{0.01998 \times 8.31 \times 298}{1 \times 10^5} = 4.95 \times 10^{-4} \text{ m}^3$$

Questions

1 Nitrogen and hydrogen react in the Haber Process to form ammonia. Calculate the volume of ammonia that will be formed when 90 cm³ of hydrogen reacts completely with nitrogen if the temperature and pressure are kept constant.

$N_2(g) + 3H_2(g) \rightarrow 2NH_3(g)$

2 Calculate the volume occupied by 2.31 moles of hydrogen gas at a pressure of 2×10^5 Pa and a temperature of 32°C.

3 Magnesium metal reacts with hydrochloric acid to form hydrogen gas.

$Mg(s) + 2HCl(aq) \rightarrow MgCl_2(aq) + H_2(g)$

Calculate the volume of gas released when 0.20 g of magnesium reacts with an excess of acid at 298 K and 1×10^5 Pa.

1.05 Writing and balancing equations

OBJECTIVES

By the end of this section you should be able to:

○ *write and balance equations for the reactions you have studied in Unit 1*

○ *balance equations for unfamiliar reactions*

○ *write an ionic equation*

○ *calculate the atom economy for a given reaction*

○ *calculate the percentage yield for a given reaction*

Acid reactions

Acids are proton donors. In their reactions their H^+ ion is displaced and a salt is formed.

You should be familiar with the following acid reactions:

acid + metal → salt + hydrogen

acid + metal oxide → salt + water

acid + metal carbonate → salt + water + carbon dioxide

acid + alkali → salt + water

Combustion reactions

All substances when burnt in air (or pure oxygen) form oxides.

e.g. calcium + oxygen → calcium oxide

Hydrocarbons when burnt in unlimited oxygen form carbon dioxide and water.

e.g. methane + oxygen → carbon dioxide + water

Balanced chemical equations

In a chemical reaction, reactants are changed into products. **Balanced equations** are written to show the reactants and products for reactions. It is important to remember that the number of each type of atom on the left hand side of the equation must be the same as the number of the same type of atom on the right hand side.

- You must follow the same steps each time you write an equation for a reaction.
- There is no need to write anything down except the final equation but you must think through the process in a logical way.

Writing a balanced equation

We can use the reaction of magnesium with oxygen as an example.

Step 1 Work out the identities of the reactants and products

$$\text{magnesium} + \text{oxygen} \rightarrow \text{magnesium oxide}$$

Step 2 Construct formulae for each of the reactants and products

$$Mg + O_2 \rightarrow MgO$$

Step 3 Balance the equation so that the number of each type of atom is the same on each side of the equation. Work from left to right balancing each atom in turn.

$$2Mg + O_2 \rightarrow 2MgO$$

Step 4 Add state symbols using (s) for solid, (l) for liquid, (g) for gas and (aq) for aqueous, a solution in water.

$$2Mg(s) + O_2(g) \rightarrow 2MgO(s)$$

In Unit 1 you will be expected to write and balance simple equations for reactions that you have studied in the unit. Examples include acid–base neutralization reactions that you carried out when completing titrations and equations for the cracking and combustion of the alkanes. You may also be given an unfamiliar equation and be asked to balance it. Do practise writing equations as you work through this revision book.

Balancing an equation for an unfamiliar reaction

Ammonia, NH_3, reacts with sodium to form sodium amide, $NaNH_2$, and hydrogen. Write a balanced equation for this reaction.

Step 1 Write formulae for the reactants on the left of an equation and for the products on the right.

$$NH_3 + Na \rightarrow NaNH_2 + H_2$$

Step 2 Work through the equation from left to right balancing each atom in turn. There is one nitrogen atom on the left and one on the right so this equation balances in terms of nitrogen.

There are three hydrogen atoms on the left and four on the right. In order to provide enough hydrogen atoms for the right we must place a two in front of the ammonia.

$$2NH_3 + Na \rightarrow NaNH_2 + H_2$$

We now need to check the nitrogen atoms again; there are two on the left so we need to place a two in front of the sodium amide.

$$2NH_3 + Na \rightarrow 2NaNH_2 + H_2$$

The nitrogen and hydrogen atoms are now balanced. To finish the equation we need to place a two in front of the sodium atom.

$$2NH_3 + 2Na \rightarrow 2NaNH_2 + H_2$$

Ionic equations

Ionic equations can be written for precipitation, displacement, and neutralization reactions. Ions which play no part in the reaction are omitted. These are called spectator ions. For the reaction of hydrochloric acid and potassium hydroxide it is possible to write both a full equation and an ionic equation.

The full equation is: $HCl(aq) + KOH(aq) \rightarrow KCl(aq) + H_2O(l)$

In this equation note that the only substance which exists as a molecule is the water; all other species are ionic. To construct an ionic equation all the ions are written out, then those that are spectator ions can be cancelled out.

$$H^+(aq) + \cancel{Cl^-(aq)} + \cancel{K^+(aq)} + OH^-(aq) \longrightarrow \cancel{K^+(aq)} + \cancel{Cl^-(aq)} + H_2O(l)$$

This gives an ionic equation of $H^+(aq) + OH^-(aq) \rightarrow H_2O(l)$

Precipitation reactions are those in which a solid substance is formed by the combination of two ions in solution. These reactions are often used to test for the presence of ions. For example, the addition of silver nitrate solution to potassium iodide solution results in the formation of a yellow precipitate of silver iodide.

The full equation is: $AgNO_3(aq) + KI(aq) \rightarrow KNO_3(aq) + AgI(s)$

Both potassium and nitrate ions remain in solution so are spectator ions.

The ionic equation is $Ag^+(aq) + I^-(aq) \rightarrow AgI(s)$

Atom economy

The **atom economy** of a chemical reaction is the proportion of reactants that are converted into useful products.

- Processes with high atom economy are more efficient and produce less waste.
- This is important for sustainable development.

Atom economy is calculated by looking at the mass of desired product in relation to the total mass of the reactants or products. These may be expressed as mass in grams or as relative molecular mass.

Calculating atom economy

a A chemical reaction produces 48 g of a desired product from reactants of total mass 75 g. Calculate the percentage atom economy.

$$\% \text{ atom economy} = \frac{\text{mass of desired product}}{\text{total mass of all reactants}} \times 100$$

$$= \frac{48}{75} \times 100 = 64\%$$

b The reaction of methane with steam produces hydrogen gas along with carbon monoxide as a waste product. Calculate the percentage atom economy in relation to hydrogen gas. $CH_4(g) + H_2O(g) \rightarrow 3H_2(g) + CO(g)$

$$\% \text{ atom economy} = \frac{\text{relative mass of desired product}}{\text{total relative mass of all products}} \times 100$$

$$= \frac{3 \times 2}{((3 \times 2) + 28)} \times 100 = \frac{6}{34} \times 100 = 17.6\%$$

Note that the ratios of each reacting species have been taken into account here.

Percentage yield

The actual yield of a product shown as a percentage of the expected yield.

$$\text{Percentage yield} = \frac{\text{actual yield}}{\text{theoretical yield}} \times 100$$

Questions

1 Ethene, C_2H_4, reacts with hydrogen bromide, HBr, to form bromoethane, C_2H_5Br. Calculate the percentage yield for this reaction if 2 g of ethene forms 5.8 g of bromoethane.

2 Chlorine gas can be obtained from the electrolysis of brine. The equation for this process is:

$2NaCl(aq) + 2H_2O(l) \rightarrow 2NaOH(aq) + Cl_2(g) + H_2(g)$.

Calculate the atom economy for producing chlorine.

3 The Haber Process for making ammonia by reacting nitrogen and hydrogen gases

$N_2(g) + 3H_2(g) \rightleftharpoons 2NH_3(g)$

typically has a percentage yield of around 15%. Explain what is meant by the term percentage yield and compare this to the atom economy for this process.

1.06 Ionic bonding

OBJECTIVES

By the end of this section you should be able to:

○ define the term ionic bond

○ draw dot and cross diagrams for ionic compounds

○ sketch the lattice structure of sodium chloride

○ construct formulae for ionic compounds

Ionic bond

An electrostatic attraction between ions of opposite charge.

Formulae of some common ions

Positive ions (cations)

hydrogen	H^+
sodium	Na^+
silver	Ag^+
potassium	K^+
lithium	Li^+
ammonium	NH_4^+
barium	Ba^{2+}
calcium	Ca^{2+}
copper(II)	Cu^{2+}
magnesium	Mg^{2+}
zinc	Zn^{2+}
lead	Pb^{2+}
iron(II)	Fe^{2+}
iron(III)	Fe^{3+}
aluminium	Al^{3+}

Negative ions (anions)

chloride	Cl^-
bromide	Br^-
fluoride	F^-
iodide	I^-
hydroxide	OH^-
nitrate	NO_3^-
oxide	O^{2-}
sulfide	S^{2-}
sulfate	SO_4^{2-}
carbonate	CO_3^{2-}

Formation of ions

All chemical bonds are forces of attraction.

Ionic bonds occur when a metal and a non-metal react to form a compound.

- Metal atoms lose electrons forming **cations**.
- For example a lithium atom loses 1 electron to form a lithium ion.

    $Li \rightarrow Li^+ + e^-$ The lithium ion has electron configuration $1s^2$.

- This ion is **isoelectronic** (has the same electron configuration) with helium.
- Non-metal atoms gain electrons forming **anions**.
- For example a fluorine atom gains 1 electron to form a fluoride ion.

    $F + e^- \rightarrow F^-$ The fluoride ion has electron configuration $1s^2\ 2s^2\ 2p^6$.

- This ion is isoelectronic with neon.

Ions are formed by the transfer of electrons from the metal atom to the non-metal atom. This is shown in a dot and cross diagram:

- Only outer shell electrons are shown in a dot and cross diagram.
- Sodium chloride is formed by the transfer of an electron from sodium to chlorine.
- The electron which has come from the sodium is shown as a dot in the dot and cross diagram.
- There is no difference between an electron shown by a dot and an electron shown by a cross. They are just symbols used to show where electrons have come from.

Dot and cross diagrams

An ionic bond is the **electrostatic attraction** between cations and anions. The ions in an ionic compound form a repeating three-dimensional structure, called a **lattice**.

- Dot and cross diagrams for an ionic bond simply show the cation and the anion.
- Think carefully about which group an atom is in to decide what the charge will be on the ion that is formed.
- Remember that when making ionic bonds atoms achieve full outer energy levels.

In these ionic compounds, different numbers of electrons are transferred depending on the group in which each atom is found.

Ionic lattices

Ionic compounds are formed of lattice structures.

- In an ionic compound each ion is surrounded by ions of opposite charge.
- For example, in the sodium chloride lattice, each chloride ion is surrounded by six sodium ions and vice versa.
- This structure is repeated throughout the ionic compound.
  - 🔊 As a result ionic compounds are said to have giant structures.

Constructing ionic formulae

Ionic compounds are not molecules.

- You cannot write a molecular formula for an ionic compound.
- Instead you write an empirical formula.
- This shows you the ratio of cations to anions present in the ionic lattice.

A formula for an ionic compound is constructed from the formulae of the cations and anions present. In most cases you can work out the charges of the cation and anion using the periodic table.

- For metal atoms the charge on the ion is the same as the group number of the metal.
- For non-metal ions the charge on the ion is the same as the group number of the atom minus 8.
- Ionic compounds have a neutral charge overall so the number of positive charges must match the number of negative charges.

For example:

- Barium chloride contains Ba^{2+} ions and Cl^- ions, so has the formula $BaCl_2$.
- Lithium oxide contains Li^+ ions and O^{2-} ions, so has the formula Li_2O.

Some ionic compounds contain compound ions or molecular ions. It is worth learning the most common of these.

For example:

- Sodium hydroxide contains Na^+ ions and OH^- ions, so has the formula NaOH.
- Sodium sulfate contains Na^+ ions and SO_4^{2-} ions, so has the formula Na_2SO_4.
- Aluminium sulfate contains Al^{3+} ions and SO_4^{2-} ions, so has the formula $Al_2(SO_4)_3$.

Sodium chloride structure

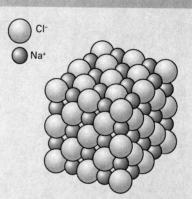

Cl^-

Na^+

A space-filling diagram showing the lattice structure of sodium chloride.

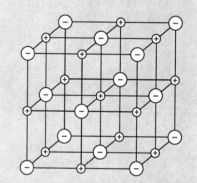

This is the easiest way to draw the lattice structure of sodium chloride. Do practise drawing this.

Questions

1 Draw a dot and cross diagram to show the ionic bond in
  - a sodium oxide
  - b magnesium chloride
  - c barium sulfide
2 Sketch the structure of sodium chloride until you can draw it from memory.
3 Construct formulae for
  - a sodium chloride
  - b magnesium bromide
  - c strontium nitrate
  - d calcium hydroxide
  - e barium carbonate

1.07 Covalent bonding

Covalent bonds

Single covalent bond: A shared pair of electrons.

Double covalent bond: Two shared pairs of electrons.

Triple covalent bond: Three shared pairs of electrons.

Questions

1 Draw dot and cross diagrams to show the bonding in:
  a Cl_2
  b CCl_4
  c PH_3
  d H_2S

2 Draw dot and cross diagrams and displayed formulae to show the bonding in
  a $NH_3 \cdot AlCl_3$
  b $PH_3 \cdot BF_3$

3 Sketch a small portion of the structures of diamond and graphite. Add labels to each diagram to explain the properties of each allotrope.

Sharing of electrons

Non-metal atoms can achieve full outer shells either by accepting electrons from metal atoms or by sharing pairs of electrons.

- **Covalent bonds** can form between identical atoms or different atoms.
- As with ionic bonding only the outer energy level electrons are involved in bonding.
- A pair of electrons is shared between two atoms.
- The bond is held together by the attraction between each nucleus involved in the bond and the pair of electrons.

Covalent bonds can be represented by dot and cross diagrams. They can also be shown as a straight line between the atoms. This is called a displayed formula.

Multiple bonds

Non-metal atoms can form double or triple covalent bonds by sharing more than one pair of electrons. These are represented in molecular formulae by multiple lines between the atoms. Double and triple bonds are stronger than single bonds.

Co-ordinate bonds

Lone pairs

Pairs of electrons that are not involved in bonding are called **lone pairs** of electrons.

- Lone pairs of electrons are important in determining the shapes of molecules.
- Lone pairs also influence some chemical properties of molecules.
- Ammonia has one lone pair of electrons since nitrogen is in group 5 and forms single covalent bonds with three hydrogen atoms.
- Water has two lone pairs since oxygen is in group 6 and forms single covalent bonds with two hydrogen atoms.

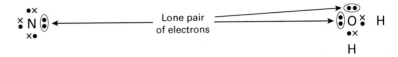

Bonding with lone pairs

Lone pairs of electrons are able to form **co-ordinate** (or dative) bonds with atoms that have vacant orbitals.

- A co-ordinate bond is a shared pair of electrons in which both electrons are contributed by one of the atoms in the bond.
- Co-ordinate bonds are shown in displayed formulae by an arrow.

For example:

- The ammonium ion, NH_4^+, has a co-ordinate bond between the nitrogen atom and one of the hydrogen atoms.
- In carbon monoxide, CO, the oxygen atom forms a double covalent bond and also a co-ordinate bond.
- In the NH_3BH_3 molecule the boron atom has a vacant orbital and accepts a pair of electrons from the nitrogen atom.

$$\left[\begin{array}{c} H \\ H\ \overset{\bullet\times}{\underset{\times\bullet}{N}}\ H \\ H \end{array}\right]^{+} \quad \left[\begin{array}{c} H \\ | \\ H-N\rightarrow H \\ | \\ H \end{array}\right]^{+} \quad \overset{\times}{\underset{\bullet}{C}}\ \overset{\times}{\underset{\bullet}{O}} \quad C\equiv O \quad \begin{array}{c} H\ \overset{\times\times}{\underset{\times\bullet}{F}}\ \\ H\ \overset{\bullet\times}{\underset{\times\bullet}{N}}\ \overset{}{B}\ \overset{\times\times}{\underset{\times\times}{F}} \\ H\ \overset{\times\times}{\underset{\times\times}{F}} \end{array} \quad \begin{array}{c} H \quad\ F \\ | \quad\ | \\ H-N\rightarrow B-F \\ | \quad\ | \\ H \quad\ F \end{array}$$

Giant covalent bonding

Molecules such as O_2, NH_3, CO_2, and H_2O are described as simple molecules. These are discrete molecules made from a small number of atoms.

- Carbon forms a number of structures called **allotropes**.
- These are different structures of the same element.
- Each carbon atom can form four covalent bonds.
- This ability to form a number of bonds enables carbon to bond to itself.
- Diamond and graphite are two different allotropes of carbon each of which has a giant molecular structure.
- A giant molecule is one in which the same arrangement of atoms is repeated many times.

In diamond:
- Each carbon atom forms four covalent bonds.
- The shape around each carbon atom is tetrahedral.
- The C–C bond angle about each atom is 109.5°.
- This tetrahedral structure is repeated around each carbon atom making diamond extremely strong.

In graphite:
- Each carbon atom forms three covalent bonds.
- The shape around each carbon atom is trigonal planar.
- The C–C bond angle about each atom is 120°.
- The carbon atoms form a flat lattice structure made from hexagons.
- The extra electron from each carbon is contributed to a delocalized sea of electrons between the layers.
- These electrons are free to move so graphite is able to conduct electricity.
- Weak van der Waals' forces exist between the layers.
- These weak forces allow the layers to slide over each other.
  - As a result graphite is soft and slippery.

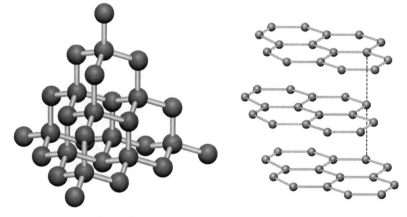

diamond   graphite

Co-ordinate bonding

Lone pair of electrons: A pair of electrons not involved in bonding.

Co-ordinate bond (or dative covalent bond): A shared pair of electrons in which one of the atoms contributes both electrons.

HSW: Diamond

This is the hardest natural substance because of its structure and bonding.

Expanding their octet

Elements in period 3 of the periodic table are able to have more than eight electrons in their outer energy level. This is because they have access to the d sub-shell which can contain ten electrons.

You are most likely to come across sulfur as an example of an element which can do this. In sulfur dioxide, for example, the sulfur atom has ten electrons in its highest energy level. Note that oxygen cannot expand beyond eight electrons as it is a period 2 element.

$$\overset{\bullet\bullet}{\underset{\bullet\bullet}{:}}O\ \overset{\times\ \times}{\underset{\times\ \times}{S}}\ O\overset{\bullet\bullet}{\underset{\bullet\bullet}{:}} \qquad \begin{array}{c} S \\ /\!/\ \backslash\backslash \\ O\quad O \end{array}$$

Note also that elements in group 3 can have vacant orbitals in their compounds.

For example: in boron trifluoride, BF_3, boron has six electrons in its highest energy level.

OBJECTIVES

By the end of this section you should be able to:

○ explain valence shell electron pair repulsion theory

○ sketch the shapes of molecules and ions containing between two and six pairs of electrons

○ recall these shapes and all of their bond angles

Valence shell electron pair repulsion theory

The shapes of molecules and ions are determined by applying valence shell electron pair repulsion theory:

- **Valence shell electrons** are those in the outermost energy level.
- These pairs of electrons repel each other.
- The shape of a molecule is such that the distance between the pairs of electrons is as large as possible.
- Multiple bonds have the same repelling effect as single bonds.
- **Lone pairs of electrons** are more repelling than bonding pairs of electrons.
  - Ⓑ As a result the bond angles are slightly smaller when lone pairs are present.

To work out the shape of a molecule you must first draw a dot and cross diagram. Remember that this diagram only includes electrons in the outermost energy level. From this diagram, you can determine the number of pairs of electrons around the central atom and hence the shape.

You must be able to remember the shapes of molecules containing between two and six pairs of electrons.

pairs of electrons	basic shape	bond angles
2	linear	180°
3	trigonal planar	120°
4	tetrahedral	109.5°
5	trigonal bipyramidal	120° and 90°
6	octahedral	90°

For molecules that are not flat a standard notation is used for showing their shape:

- The dotted line shows a bond going behind the plane of the paper.
- The shaded triangle shows a bond coming out of the plane of the paper.
- The ordinary line shows a bond in the plane of the paper.

It is always advisable to sketch and name a shape in an examination. This helps if your drawing skills are not particularly strong!

Molecules containing one or more lone pairs of electrons

Tetrahedral

Both ammonia, NH_3, and water, H_2O, are molecules containing four pairs of electrons overall.

- In ammonia one of these pairs of electrons is a lone pair.
  - Ⓑ As a result ammonia has a pyramidal shape.
- In water two of the pairs of electrons are lone pairs.
  - Ⓑ As a result water has a bent shape.
- The repulsive effect of a lone pair of electrons is greater than a bond pair.
  - Ⓑ As a result the bond angles in ammonia and water are slightly smaller than that of methane.

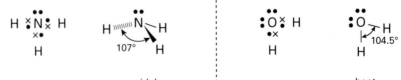

pyramidal   bent

Molecular Shapes

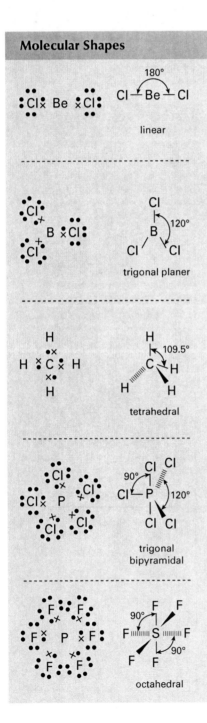

linear

trigonal planer

tetrahedral

trigonal bipyramidal

octahedral

Trigonal bipyramidal

Sulfur tetrafluoride, SF_4 and chlorine trifluoride, ClF_3 both have five pairs of electrons around the central atom.

- In sulfur tetrafluoride one of these pairs of electrons is a lone pair.
- So one corner of the trigonal bipyramid is occupied by the lone pair.
- In chlorine trifluoride there are two lone pairs of electrons.
- Two corners of the trigonal bipyramid are occupied by lone pairs.
  - As a result chlorine trifluoride is T-shaped.

Octahedral

Iodine pentafluoride has six pairs of electrons around the central atom.

- One of these pairs of electrons is a lone pair.
- So one corner of the octahedron is occupied by the lone pair.
  - As a result iodine pentafluoride is square pyramidal.

Shapes of ions

The shapes of molecular ions can be worked out in the same way as for uncharged molecules.

- Remember to take account of the charge on the ion when working out the dot and cross diagram.
- Positively charged ions have fewer electrons than the original atom.
- Negatively charged ions have gained extra electrons. These can be indicated on a dot and cross diagram using a triangle.

Examine the diagrams for the shapes of ions carefully. Note the following points:

- in the NH_4^+ molecule there is a dative covalent bond between the nitrogen and one of the hydrogen atoms.
- in I_3^- there is an extra electron due to the negative charge shown with a triangle. Note that iodine can have more than eight electrons in its highest energy level as it is in period 4. The shape is based on a trigonal bipyramid with three lone pairs.
- in ICl_4^- the lone pairs occupy the vertical positions of the octahedron giving a square planar shape.

The sulfate ion, SO_4^{2-} has a complex dot and cross diagram:

- The sulfur atom forms two double bonds and two single bonds.
- The two electrons from the two minus negative charge are placed on the singly bonded oxygen atom in order to give the oxygen a complete outer energy level.
- The electrons in the S=O bonds are delocalized (spread out) over the structure.
  - As a result the shape of the molecule is tetrahedral with equal bond angles of $109.5°$.

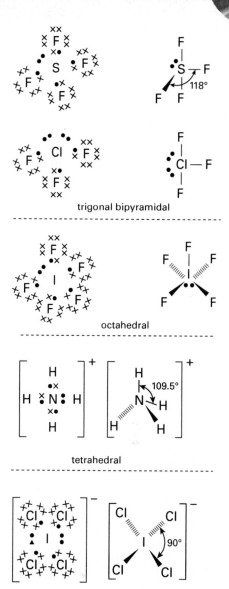

trigonal bipyramidal

octahedral

tetrahedral

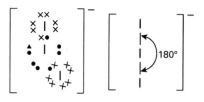

square planer

linear

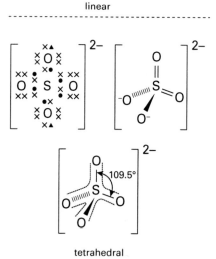

tetrahedral

Questions

1 Draw a dot and cross diagram for each of the following molecules. Then use it to draw a labelled diagram of the molecule showing all bond angles.

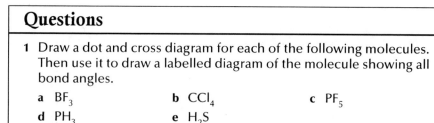

   a BF_3 **b** CCl_4 **c** PF_5

   d PH_3 **e** H_2S

2 Draw a dot and cross diagram for the following ions. Remember to take the charges into account. Use the dot and cross diagram to draw each ion.

   a BF_4^- **b** ICl_4^+ **c** PF_6^-

Electronegativity, polar bonds, and dipoles

Electronegativity: The ability of an atom to withdraw electron density from a covalent bond.

Polar bond: A covalent bond between atoms with different electronegativities.

Dipole: Opposite charges separated by a short distance in a molecule or ion.

HSW: Linus Pauling

Linus Pauling was awarded the Nobel Prize in Chemistry in 1954 for his research into the nature of chemical bonds. In 1962 he was awarded the Nobel Peace Prize for his efforts to stop nuclear weapons testing.

Charge distribution in covalent molecules

Covalent bonding is the sharing of one or more pairs of electrons.

The covalent bond is held together by the attraction between the nuclei of the two atoms involved in the bond and the pairs of electrons.

- If the bond is between identical atoms this sharing is equal e.g. H_2, Cl_2.
- If the atoms are different then the sharing may be unequal e.g. H–F.
- The fluorine atom is much better at attracting the pair of electrons than hydrogen because it has a greater number of protons.

Electronegativity

Electronegativity is the ability of an atom to withdraw electron density from a covalent bond.

- If the two atoms in a bond have different electronegativites then the more electronegative element has a greater share of the electrons.
- The electronegativity of an element can be calculated.
- The most common scale for electronegativity is the Pauling electronegativity scale.

You do not need to remember this scale but do need to understand that non-metal elements at the top of groups 5, 6, and 7 are the most electronegative.

Electronegativity trends

There are two main trends in electronegativity:

- Electronegativity increases across the periodic table as the number of protons in the nucleus increases. This increases the ability of an atom to attract electrons.
- Electronegativity decreases down the periodic table as the amount of electrons in complete energy levels increases. As a result the nucleus is shielded and has less ability to attract electrons.
- Fluorine is the most electronegative element owing to its small size.

Polar bonds

Polar bonds are those in which the pair of electrons is not shared equally.

- The more electronegative element has a partial negative charge, shown by $\delta-$.
- The less electronegative element has a partial positive charge, shown by $\delta+$.

In the hydrogen chloride molecule:

- The chlorine atom is more electronegative so has a partial negative charge.
- The hydrogen atom therefore has a partial positive charge.
  ⊗ As a result, The H–Cl *molecule* can be described as polar. $\overset{\delta+}{H}-\overset{\delta-}{Cl}$

In the carbon dioxide molecule:

- The oxygen atoms are more electronegative so have a partial negative charge.
- The carbon atom has two partial positive charges as it is bonded to two oxygen atoms.
  ⊗ As a result, The C=O *bond* can be described as polar. $\overset{\delta-}{O}=\overset{2\delta+}{C}=\overset{\delta-}{O}$

The Pauling scale for electronegativity

H 2.2							He –
Li 1.0	Be 1.6	B 2.0	C 2.5	N 3.0	O 3.4	F 4.0	Ne –
Na 0.9	Mg 1.3	Al 1.6	Si 1.9	P 2.2	S 2.6	Cl 3.2	Ar –
K 0.8	Ca 1.0					Br 3.0	Kr 3.0
Rb 0.8						I 2.7	Xe 2.6

Polar bonds and polar molecules

A molecule that contains polar bonds may not be a polar molecule.

To find out if a molecule is polar:

- Draw the molecule (in three dimensions if necessary, remembering about the influence of lone pairs).
- Label any polar bonds using the δ+ , δ– convention.
- Then examine the shape of the molecule.
- If the polar bonds cancel out then the molecule is not polar.
- If the polar bonds do not cancel out then the molecule is polar.

Carbon dioxide has two polar bonds but there is no net dipole as the bonds are symmetrical. Carbon dioxide is a non-polar molecule.

Boron trifluoride has three polar bonds but there is no net dipole as the bonds are symmetrical. Boron trifluoride is a non-polar molecule.

Water has two polar bonds and has a net dipole beacause the polar bonds do not cancel each other out. Water is polar molecule. This has a big influence on the properties of water.

The character of bonds

Covalent bonds

The greater the electronegativity difference between two atoms the bigger the **dipole** (difference in charge between the atoms) and the more polar the bond.

For example:

- The H–F bond has an electronegativity difference of 1.8.
- The H–Cl bond has an electronegativity difference of 1.0.
- The H–Br bond has an electronegativity difference of 0.8.
  - ⊗ As a result the H–F bond is the most polar

If there is a large difference in electronegativity, the covalent bond can be described as having ionic character. So H–F has the most ionic character.

- If there is a very large difference in electronegativity then the bonding is ionic rather than covalent.
- It is a good idea to think of ionic and covalent bonding as the extreme ends of a range of types of bonding.

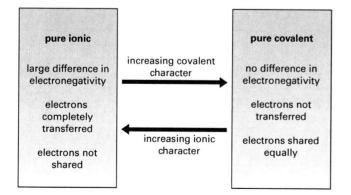

pure ionic		pure covalent
large difference in electronegativity	increasing covalent character →	no difference in electronegativity
electrons completely transferred		electrons not transferred
electrons not shared	← increasing ionic character	electrons shared equally

Ionic bonds

Ionic bonds can also show polarity.

- The electron cloud around a large negative ion can be distorted by a small highly charged positive ion (these have high charge density).
- The positive ion is said to be **polarizing.**
- The negative ion is **polarized.**
- If this happens to a large enough extent the ionic bond takes on covalent character.

For example:

- Sodium ions, Na^+, have a low charge density.
- This means they are not able to polarize chloride ions.
- Sodium chloride has ionic bonding.
- Aluminium ions, Al^{3+}, have a high charge density.
- This means they are able to polarize chloride ions.
- This polarization is so great that aluminium chloride has covalent bonding.

Questions

1 Label the following bonds to indicate their polarity.

   a H–F

   b C–Cl

   c O–N

   d S=O

2 Draw the following molecules. Label any polar bonds and state whether the molecule is polar.

   a hydrogen bromide

   b hydrogen sulfide

   c ammonia

   d fluorine oxide, F_2O

3 Arrange the substances below in order of their covalent character starting with the least covalent:

   a $NaCl$, $AlCl_3$, $MgCl_2$

   b NaI, $NaCl$, $NaBr$

1.10 Intermolecular forces 1

OBJECTIVES

By the end of this section you should be able to:

O explain what is meant by an intermolecular force

O describe a permanent dipole–dipole force

O define a hydrogen bond

O draw a diagram to show the hydrogen bonding in water and ice

O explain the influence of hydrogen bonding on physical properties

Intermolecular forces

An intermolecular force is a weak attractive force between molecules.

A permanent dipole–dipole attraction is an attractive force that exists between polar molecules.

A hydrogen bond is an intermolecular force between a lone pair of electrons on an N, O, or F atom in one molecule, and an H atom joined to an N, O, or F atom in another molecule.

Intermolecular forces are weak attractions that exist between molecules. These are the forces that hold substances together in the solid or liquid form. The strength of these forces determines the melting points and boiling points of substances and can influence some of their other properties too.

Permanent dipole–dipole forces

Molecules with a **permanent dipole** have regions of different electron density within them. These molecules are described as being polar.

It is easy to see that if two polar molecules come near each other in space there will be an attraction between them.

For the hydrogen chloride molecule below, the electronegative chlorine of one HCl molecule will attract the electropositive hydrogen of another.

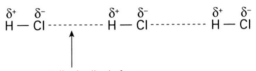

permanent dipole-dipole force

- In liquid HCl these attractions are constantly breaking and re-forming as the molecules move around slowly.
- In solid HCl these attractions hold the molecules in a fixed position.
- This has a big influence on the boiling point of hydrogen chloride.

Forces of this type exist between any two molecules that have permanent dipoles.

Hydrogen bonding

Hydrogen bonds are an especially strong **permanent dipole–dipole force**, which exists between molecules that contain very electronegative elements.

For a hydrogen bond to occur:

- The molecule must have an O, N, or F atom bonded to a hydrogen atom.
- The molecule to which it is attracted must contain an O, N, or F atom.
- O, N, and F are the most electronegative atoms.
  - As a result the dipole–dipole force between the molecules is especially strong.

The most common examples of hydrogen bonding asked about in AS exams are

- water
- ammonia
- hydrogen fluoride

There are, of course, many other molecules that are capable of hydrogen bonding and you will meet some of these in A2 and in other science courses. For example, protein chains are linked together by hydrogen bonds.

When drawing a diagram to show hydrogen bonding you must include the following:

- labelled dipoles on every molecule
- lone pairs on the O, N, or F atom
- a dotted line to represent the hydrogen bond

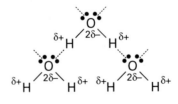

Hydrogen bonding in water.

Hydrogen bonding and physical properties

Intermolecular forces have a big influence on physical properties. These are properties such as melting point, boiling point, density, etc.

The two boiling point graphs show the significantly larger boiling points for water and hydrogen fluoride in comparison to the hydrides of other members of the same group.

We will revisit these graphs in the next section to consider the rest of the pattern.

The boiling points of the group 6 hydrides

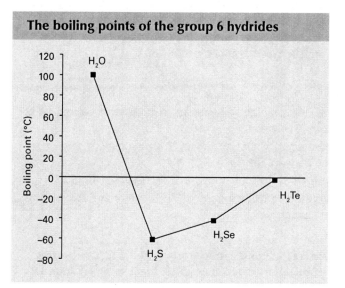

The boiling points of the group 7 hydrides

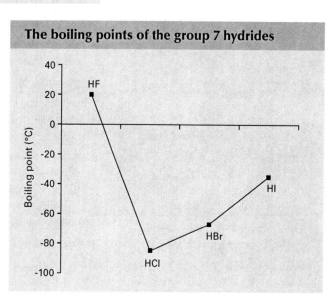

Ice

Ice has a very regular structure held together by hydrogen bonding between molecules. This hydrogen bonding causes water to have some unusual properties.

- Ice floats on water. It is the only substance for which this is the case.
- Water has a large surface tension, caused by a network of hydrogen bonds on the surface.
- Water has a higher boiling point than would be expected. This is because hydrogen bonds must be broken when water is boiled.

The structure of ice

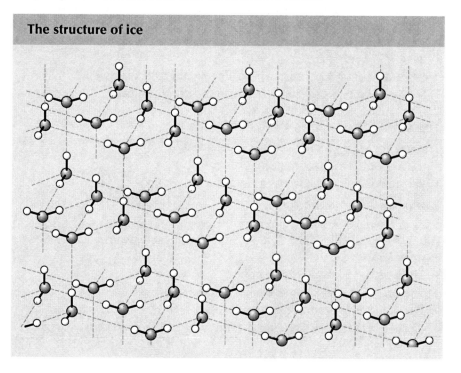

Questions

1 Draw a labelled diagram from memory to show the hydrogen bonding in water. Check your diagram against the book. Make sure that you have labelled the dipoles, shown lone pairs, and labelled your dotted line.

2 Draw a labelled diagram to show the hydrogen bonding in

   a hydrogen fluoride

   b ammonia

3 Make a list of the unusual properties of water. Write a one sentence explanation of each one.

Temporary dipole–induced dipole force

These forces are also known as van der Waals' forces

Temporary dipole: The asymmetrical distribution of the electron pair in a covalent bond.

Induced dipole: An uneven distribution of charge in a molecule or atom, caused by a charge in an adjacent particle.

The formation of van der Waals' forces

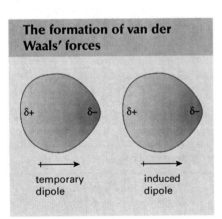

temporary dipole

induced dipole

The structure of iodine

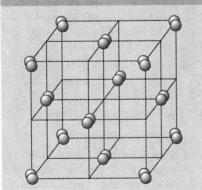

Van der Waals' forces

Molecules without permanent dipoles can form liquids and solids so there must be **intermolecular forces** between the molecules. Remember that it is intermolecular forces that are broken on melting or boiling, *not* covalent bonds.

HSW: *The name given to this force is a **van der Waals' force** after the Dutch scientist Johannes van der Waals who won the Nobel Prize in Physics in 1910.*

- A van der Waals' force is a force of attraction between a temporary dipole on one molecule and an induced dipole on another molecule.
- Van der Waals' forces do form between polar and non-polar molecules but are the dominant force between non-polar molecules.

Temporary dipoles

Electrons in a molecule are constantly moving.

- This means that the electron cloud around an atom or within a non-polar molecule is not static.
- At any instant in time the distribution of the electrons may be uneven although on average they are distributed evenly.
  - ® As a result a non-polar molecule may have a temporary dipole.

The dipoles can be represented using an arrow. The head of the arrow shows the region of negative charge.

Induced dipoles

The presence of a **temporary dipole** in one atom or molecule can cause a dipole to form in a nearby atom or molecule. This dipole is called an **induced dipole**. The induced dipole can then induce a dipole in a neighbouring atom or molecule. The net effect of this is a force of attraction between the particles called a **temporary dipole–induced dipole force**.

Van der Waals' forces and structure

All non-polar atoms or molecules have van der Waals' forces between the particles when they are in the liquid or solid state.

For example, iodine is a group 7 element that exists as diatomic molecules (two atoms joined together with a covalent bond).

- Iodine is a crystalline solid at room temperature.
- So the molecules in the iodine are in a regular arrangement, with van der Waals' forces between the molecules.
- These forces are relatively weak so little energy is needed to overcome them.
  - ® As a result iodine has a low melting point and boiling point.
- Iodine can sublime (change directly from a solid to a gas).

The strength of van der Waals' forces

The van der Waals' force strength is dependent on

- the size of the atom or molecule
- the area of contact between the atoms or molecules

Size of the atom or molecule

Look at the trend in boiling points of the group 7 elements shown in the graph. Work carefully through these points, which often comes up in AS examinations.

- The boiling points increase significantly down group 7.
- The size of the halogen molecules increases down group 7.
- The number of electrons in each molecule increases down the group.
  - ® As a result temporary dipoles form more readily in the large halogen molecules.

- In addition dipoles are more readily induced in adjacent molecules.
  As a result the van der Waals' forces get stronger down the group.

The same trend occurs in the alkanes as the number of carbon atoms in the chain increases.

The boiling points of the alkanes

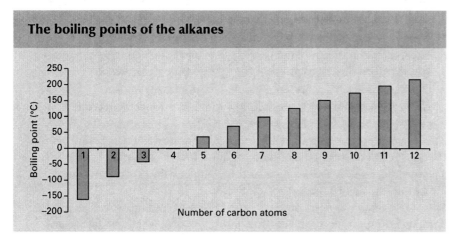

The boiling points of the halogens

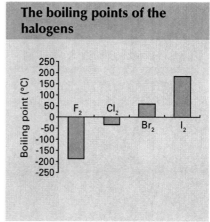

Area of contact between molecules

The table of data below shows the boiling points of three alkanes. These all have the same molecular formula C_5H_{12} but have different chain lengths (you will revisit this work in the section on alkanes).

name	displayed formula	boiling point (°C)
pentane		36
2-methylbutane		28
2,2-dimethylpropane		10

The molecule with the lowest boiling point has the most chain branching. This increase in chain branching reduces the area of contact with other molecules and therefore reduces the strength of the van der Waals' force. Note – you do not need to remember this data but do need to remember the trend!

Questions

1 One of the first substances studied by Johannes van der Waals in the late 1800s was argon.

 a Complete the table below with your prediction for the melting point of argon.

 b Then write a one sentence explanation of your prediction.

element	atomic mass	melting point (°C)
helium	4	−272
neon	20	−249
argon	40	
krypton	84	−156
xenon	132	−112

2 Practise sketching the structure of solid iodine. Label your diagram clearly to show the intermolecular force.

3 Think about the boiling of bromine. Write a short paragraph explaining what happens to the molecules of bromine as it forms bromine gas.

Metallic bonding

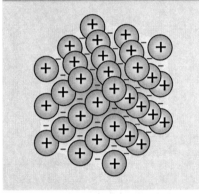

Metallic bond

Metallic bond: The electrostatic force of attraction between metal ions and the delocalized electrons in a metallic lattice.

A

layer one

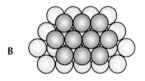

B

layer two

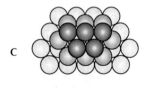

C

layer three

'Building' an HCP structure.

The metallic bond

Atoms of metals are held together by **metallic bonds**. In a metallic bond:

• Each metal atom forms a positive ion (cation).

• The positive ions are arranged into a regular **lattice** structure.

• The ions in the structure are very close to each other so the electrons that are lost when the metal atoms form ions become **delocalized**.

• Delocalized electrons are not attracted to any particular ion.

  ⊛ As a result these electrons are free to move through the metal.

For example, magnesium atoms have the electron configuration $1s^2\ 2s^2\ 2p^6\ 3s^2$. The 3s electrons are delocalized from each atom forming magnesium ions of charge 2^+. These have the electron configuration $1s^2\ 2s^2\ 2p^6$.

The metallic bond is the strong attraction between the positive ions in the lattice and the delocalized sea of electrons.

Note – although the term lattice is used for ionic bonding as well as metallic bonding you must be careful not to confuse the two. The ionic lattice is a regular arrangement of ions of opposite charge. The metallic lattice is a regular arrangement of cations surrounded by a sea of delocalized electrons.

Metallic bond strength

Metallic bonds do not all have the same strength. If they did, then all metals would melt and boil at the same temperature.

The strength of a metallic bond is dependent on

• the charge of the ions in the lattice

• the number of electrons in the sea of delocalized electrons

The boiling point graph for sodium, magnesium, and aluminium shows this. Magnesium needs more energy to boil than sodium as the attraction between the Mg^{2+} ions in the lattice and the sea of electrons is greater than between the Na^+ and the sea. The force of attraction is even stronger in aluminium which has Al^{3+} ions.

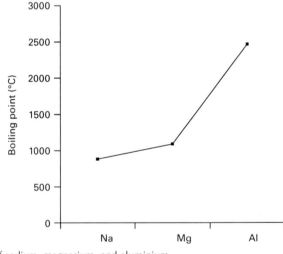

Boiling points of sodium, magnesium, and aluminium

The structure of magnesium

The cations formed in metallic bonding arrange themselves in a **close-packed structure**.

• There are two forms of close-packed structure: hexagonal close-packed (HCP) and simple cubic.

• In the simple cubic structure the ions are stacked directly on top of each other.

• The hexagonal close-packed structure is more efficient, with 74% of the volume filled with particles. (See left)

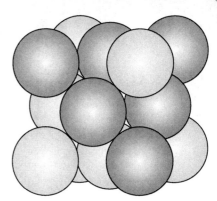

1 The ions in the first layer are arranged so that they are touching each other.
2 In the second layer the ions sit in the hollows between the ions in the first layer.
3 In the third layer the ions are in hollows in the second layer, directly over the ions in the first layer.

You might like to try building this structure with some marbles (but put them into a tray first!).

This hexagonal close-packed structure can be shown as either a space-filling model or a unit cell.

Properties of metals

The properties of metals can be easily described in terms of their lattice structure.

- Metals are **good conductors** of heat and electricity.
- Metals have high melting points and boiling points.
- Metals are **malleable**.
- Metals are **ductile**.

Conducting heat and electricity

Metals are good conductors of heat as the metal ions in the lattice are very close to each other. Heating one end of a piece of metal makes the ions at that end vibrate more, and these vibrations are passed along the piece of metal.

Metals are good electrical conductors as the sea of delocalized electrons is free to move when a voltage is applied. Remember that the electrons are not attracted to any particular ion. 'Pushing' electrons into one end of the piece of metal causes electrons to come out of the other end.

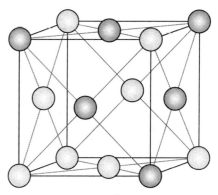

An HCP unit cell.

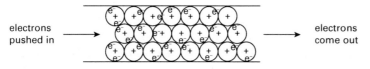

electrons
pushed in → ← electrons
come out

Malleability and ductility

Malleability is the ability of a piece of metal to be pressed or hammered into shape. Ductility is the ability to be stretched into a wire.

Both of these properties are due to the regular layer structure of the metal. The layers of cations can slip over each other if a large enough force is applied. The strength of the metallic bond stops the attractions being broken completely.

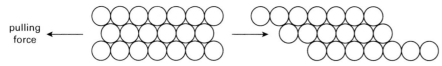

pulling
force ←

Questions

1 Draw a simple sketch of the structure of a piece of aluminium. Label your sketch to show how aluminium is able to conduct electricity.
2 List the following metals in order of their melting point. Explain the order you have chosen: sodium, aluminium, magnesium.
3 List the key properties of metals and write a two sentence explanation of each property in terms of the structure of the metal.

1.13 Structure and bonding

OBJECTIVES

By the end of this section you should be able to:

○ explain the difference in particle arrangements between solids, liquids, and gases

○ explain the process of changing state

○ identify the force being broken when substances are changing state

○ describe why some substances are able to conduct electricity

Particle diagrams

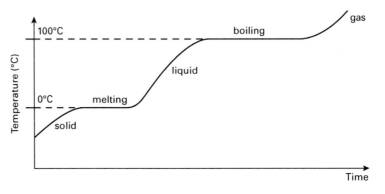

fixed position regular lactice arrangement — solid — particles close together

particles moving around random arrangement — liquid

gas — particles far apart

Changing state

When a solid is melted or a liquid frozen, a liquid boiled or a gas condensed it is said to have **changed state**.

- All changes of state involve changes in energy.
- When a solid melts or a liquid boils, the energy is used to break the forces between the atoms, molecules, or ions involved.
- As the change of state occurs, the temperature stays constant because the energy provided to the system is used to break the force.

It is essential that you are able to identify the type of force being broken in each case. Read this section after you have worked through the pages on ionic bonding, covalent bonding, metallic bonding, and intermolecular forces.

Metals

When melting a pure metal the attraction between the **lattice** of positive ions and the **delocalized sea** of electrons is broken. This is a strong force so requires a large amount of energy, which means that metals have high melting points. The melting points increase as the charge on the metal ion and the number of electrons in the sea increases.

Ionic substances

Remember that an ionic bond is an **electrostatic attraction** between ions of opposite charge. When an ionic substance melts, the energy provided is used to break this attraction. The attraction is strong so ionic substances are hard to melt.

Simple molecular substances

Melting simple molecular substances requires the breaking of the **intermolecular force** between the molecules.

- There are three types of forces that can be broken.
- Each of these forces is much weaker than the covalent bond that exists between the atoms within the molecule.

You must be able to identify the type of force between molecules and relate this to changes of state.

name of force	requirement for force	example of molecule	relative strength of force
permanent dipole–dipole attraction	polar molecule	HCl PH$_3$	stronger than van der Waals' force, weaker than hydrogen bond
hydrogen bond	O–H, F–H, N–H bond present in molecule	H$_2$O HF NH$_3$	the strongest intermolecular force
van der Waals' force	non-polar molecule	Cl$_2$, Br$_2$, I$_2$, CH$_4$, C$_2$H$_6$, etc.	weakest intermolecular force

Giant molecular substances

Giant molecular substances have a large network of covalent bonds. Melting these substances involves a large amount of energy as strong covalent bonds must be broken in order to change state. You must be able to explain why diamond and graphite have such high melting points.

Conducting electricity

In order to conduct electricity charged particles must be able to move through a substance when a voltage is applied:

- In a metal the delocalized electrons are free to move.
- In a molten ionic substance the ions are free to move. Note that ionic solids do not conduct electricity as the ions are held in a fixed position.
- Graphite is able to conduct as there are delocalized electrons between the layers.

The structure of graphite

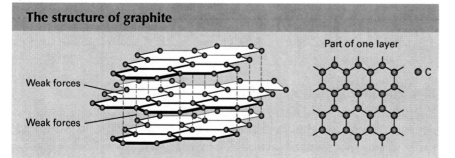

Part of one layer

Weak forces

Weak forces

● C

Polarizing ions

Remember that ionic bonds also have the ability to show some polarity.

- The electron cloud around an ion can be distorted by the presence of a positive ion.
- The positive ion is said to be **polarizing**.
- The negative ion has been **polarized**.
- If the degree of polarization is large enough the ionic bond has some covalent character.

The extent to which polarization occurs is dependent on the size and charge of the ions.

- Cations are most polarizing if they are small and highly charged.
- Anions are most readily polarized if they are large and have a high charge (this gives them a high charge density).

A comparison of sodium chloride and aluminium chloride puts this theory into practice. In sodium chloride the bonding is ionic as sodium ions have a low charge density and cannot polarize chloride ions. Aluminium chloride has covalent bonding as Al^{3+} ions are so small and highly charged they have a high charge density and have a very large polarizing effect on the chloride ions.

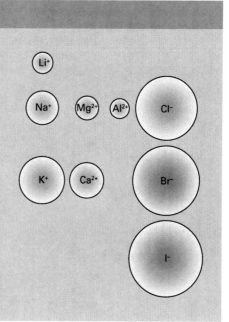

Questions

1 Complete the table below showing the type of bonding in each substance

name of substance	formula	type of structure	type of bonding
magnesium	Mg	giant	
sodium chloride			ionic
chlorine			
graphite			

2 Sketch a small section of diamond and of graphite. Use your diagrams to explain why graphite is able to conduct electricity and diamond is not.

3 Arrange the following substances in order of melting point starting with the lowest. Explain the order you have chosen. H_2O, $MgCl_2$, Cl_2.

First ionization energy

First ionization energy is the energy required to remove 1 electron from each atom in 1 mole of gaseous atoms forming 1 mole of ions with a single positive charge.

First ionization energy for period 3 elements

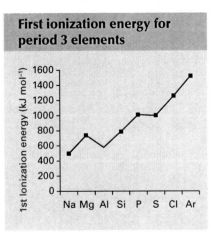

This section draws together lots of the content on structure and bonding and applies it to examining the trends in period 3 elements. Periodic chemistry is very different from group chemistry as the elements change from metallic to non-metallic across a period.

Atomic radius

The **atomic radius** of the elements decreases across period 3. Think carefully about this trend. Remember that the number of protons in the nucleus increases across the period and that the extra electrons are placed into the same shell. This means that the effect of the nucleus on the electrons increases slightly across the period, making the atomic radius slightly smaller.

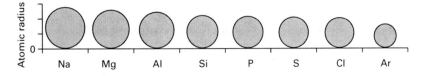

Electron configurations and first ionization energy

You should be able to write the electron configuration for all the atoms in period 3. Practise reading this information from the periodic table. Remember that elements with outer electrons in the s energy level can be classified as s block elements and those with outer electrons in the p energy level can be classified as p block elements.

symbol	type of element	electron configuration	block in the periodic table
Na	metal	$1s^2\,2s^2\,2p^6\,3s^1$	s block
Mg	metal	$1s^2\,2s^2\,2p^6\,3s^2$	s block
Al	metal	$1s^2\,2s^2\,2p^6\,3s^2\,3p^1$	p block
Si	non-metal	$1s^2\,2s^2\,2p^6\,3s^2\,3p^2$	p block
S	non-metal	$1s^2\,2s^2\,2p^6\,3s^2\,3p^3$	p block
P	non-metal	$1s^2\,2s^2\,2p^6\,3s^2\,3p^4$	p block
Cl	non-metal	$1s^2\,2s^2\,2p^6\,3s^2\,3p^5$	p block
Ar	non-metal	$1s^2\,2s^2\,2p^6\,3s^2\,3p^6$	p block

It is essential that you can sketch and explain the graph showing the **first ionization energy** for the elements in period 3. The key details are:

- There is a general increase across period 3 as the number of protons in the nucleus steadily increases, leading to a greater charge on the nucleus.
  - Ⓔ As a result there is a greater attraction between the nucleus and the outer electron.
- There is a slight decrease between magnesium and aluminium because the outer electron in aluminium is in a p sub-level which is higher in energy than the outer electron in magnesium which is in an s sub-level.
- There is a slight decrease between phosphorus and sulfur. The outer electron configuration of sulfur has a pair of electrons in the p_x orbital. There is repulsion between the electrons so less energy is needed to remove an electron from this pair.

element	electron configuration	spin diagrams					
		1s	2s	2p	3s	3p
P	$1s^2\,2s^2\,2p^6\,3s^2\,3p^3$	↑↓	↑↓	↑↓ ↑↓ ↑↓	↑↓	↑ ↑ ↑
S	$1s^2\,2s^2\,2p^6\,3s^2\,3p^4$	↑↓	↑↓	↑↓ ↑↓ ↑↓	↑↓	↑↓ ↑ ↑

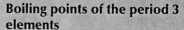

Melting points and boiling points

Trends in melting point and boiling point across period 3 are related to the change in bonding from metallic to covalent as you move across the period. It is important that you can

- state the type of bonding in each element
- sketch the structure of each element
- sketch and explain the melting point and boiling point graph.

Boiling points of the period 3 elements

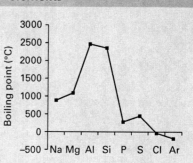

element	type of bonding	force broken on melting
sodium	metallic	metallic bond
magnesium	metallic	metallic bond
aluminium	metallic	metallic bond
silicon	giant covalent	covalent bond
phosphorus	simple covalent, P_4 units	van der Waals' force
sulfur	simple covalent, S_8 units	van der Waals' force
chlorine	simple covalent, Cl_2 units	van der Waals' force
argon	monatomic	van der Waals' force

Structures of some period 3 elements

Phosphorus, sulfur, and chlorine are all simple covalent molecules with **van der Waals' forces** between them. This force strength increases with the increasing size of the molecule so the melting or boiling point increases in the order argon, chlorine, phosphorus, and sulfur.

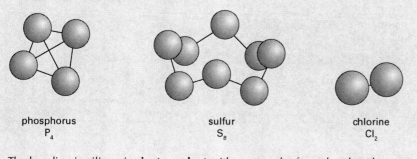

phosphorus
P_4

sulfur
S_8

chlorine
Cl_2

The bonding in silicon is **giant covalent** with a network of covalent bonds throughout the structure. Melting or boiling silicon requires the breaking of covalent bonds so requires a lot of energy. It is worth practising drawing a small section of silicon using shaded and dotted lines to show the tetrahedral structure about each silicon atom (which is exactly the same as diamond).

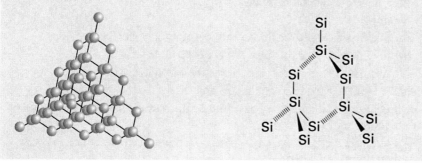

Questions

1 Practise sketching the curve for first ionization energy for period 3 elements. Cover up small sections of it then try and fill them in.

2 Draw an electron spin diagram for magnesium and aluminium. Use the diagram to explain the decrease in first ionization energy between magnesium and aluminium.

3 Draw the structures of silicon, sulfur, and phosphorus. Use them to describe the differences in boiling point.

OBJECTIVES

By the end of this section you should be able to:

○ understand the terms functional group and homologous series

○ write and interpret empirical formulae, molecular formulae, structural formulae, and displayed formulae

○ be able to apply IUPAC rules for naming alkanes, alkenes and haloalkanes with up to six carbon atoms

Key terms

Homologous series: A group of compounds with the same general formula.

Empirical formula: The simplest whole number ratio of the atoms in a compound.

Molecular formula: The actual number of each type of atom in a compound.

Displayed formula: A formula showing all the bonds in a compound.

Structural formula: A formula showing the atoms present and the bonds between them.

Questions

1 Name the following compounds:

  a C_6H_{14}

  b C_5H_{10}

  c $CH_3CHCHCH_3$

  d $CH_3CH_2CH(CH_2CH_3)CH_2CH_2CH_3$

  e $CH_3CHClCH_2CH_3$

  f $CH_3CH_2CH(CH_3)CHClCH_3$

2 Write out the displayed formula, molecular formula, and structural formula of

  a pentane

  b 2,3-dimethylbutane

  c pent-2-ene

  d 3-iodohexane

  e 2,4-dichloropentane

The chemistry of carbon

Organic chemistry is the study of compounds containing carbon combined with other elements. Carbon is an unusual element, which has the ability to form chain and ring structures by joining carbon atoms together. This ability means that a very large number of organic compounds exist.

Organic compounds are grouped into 'families' which have the same molecular formula. Each of these families is called a **homologous series**.

For Unit 1 you need to be familiar with three different homologous series:

- alkanes
- alkenes
- haloalkanes

Naming organic molecules

Functional groups

functional group	name
alkane	-ane
alkene	-ene
haloalkane	bromo-
	chloro-
	iodo-

Number of carbon atoms

number of C atoms	prefix
1	meth
2	eth
3	prop
4	but
5	pent
6	hex

Side chains

number of C atoms	structure	Name
1	$-CH_3$	methyl
2	$-CH_2CH_3$	ethyl
3	$-CH_2CH_2CH_3$	propyl

Number of identical side chains

number of identical side chains	prefix
2	di
3	tri
4	tetra

The alkanes

The **alkanes** are **hydrocarbons**. This means they only contain carbon and hydrogen atoms.

They have single covalent bonds between the carbon atoms.

Alkanes have the general formula C_nH_{2n+2}

For example:

- The alkane with only 1 carbon atom has the formula CH_4.
- The alkane with 2 carbon atoms has the formula C_2H_6.
- The alkane with 26 carbon atoms has the formula $C_{26}H_{54}$.

The alkanes all have names ending in -ane.

For straight chain alkanes the beginning of the name tells you the number of carbon atoms in the chain.

- For example methane has one carbon atom in the chain; ethane has two carbon atoms, and so on.

If the chain is branched then it is said to have a side chain. The side chain is named methyl, ethyl, propyl, etc. depending on the number of carbon atoms it contains. This is then put at the front of the name of the main chain.

- For example methylbutane has a main chain of four carbon atoms and a side chain of one carbon atom.

If there is more than one position that the side chain could attach to the main chain then it is given a number. This is counted from the end of the chain in order to make the number of the side chain as small as possible.

- For the methylbutane molecule below a 2 is placed in front of the methyl group making 2-methylbutane.

If more than one side group of the same kind is attached to the chain then the prefix di-, tri-, etc. is used in front of the name of the side chain.

For each of the alkane molecules three types of formulae have been given: **displayed**, **structural**, and **molecular**. Make sure you are confident about what each of these show and that you can write and interpret each one.

The alkenes

The **alkenes** are also hydrocarbons, like the alkanes.

They have a double covalent bond between two of the carbon atoms. Alkenes have the general formula C_nH_{2n}

For example:

- The alkene with 2 carbon atoms has the formula C_2H_4.
- The alkene with 3 carbon atoms has the formula C_3H_6.
- The alkene with 12 carbon atoms has the formula $C_{12}H_{24}$.

The alkenes are named in the same way as the alkanes except that their name ends in -ene. The position of the C=C double bond is numbered if there is more than one place where it can go in a molecule. The numbering is such that it has the smallest number possible.

The haloalkanes

The **haloalkanes** are compounds containing carbon, hydrogen, and a halogen (fluorine, chlorine, bromine, or iodine) atom.

Haloalkanes have the general formula $C_nH_{2n+1}X$ where X is the halogen atom.

For example:

- Chloromethane is CH_3Cl since it has one chlorine atom and one carbon atom.
- Bromoethane is C_2H_5Br as it has one bromine atom and two carbon atoms.

If the halogen has more than one of the same halogen atom then the halogen is numbered di-, tri-, tetra-, etc. as with the side chains in alkanes.

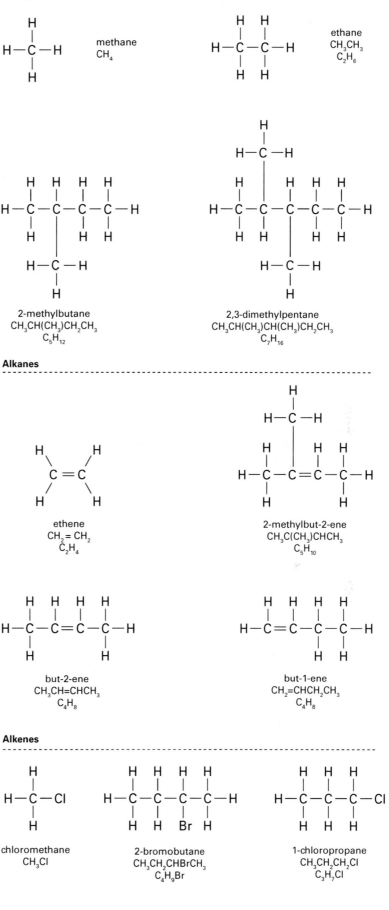

Alkanes

Alkenes

Haloalkanes

By the end of this section you should be able to:

○ *understand the difference between chain and positional isomers*

○ *draw displayed formulae and name chain and positional isomers of alkanes, alkenes, and haloalkanes*

○ *determine the molecular formula of a compound from percentage composition data and its mass spectrum*

Chain and positional isomers

Chain isomers: Molecules with the same molecular formula but different arrangements of the carbon chain.

Positional isomers: Molecules with the same molecular formula but with the functional group at different positions on the carbon chain.

Isomerism occurs when two or more organic molecules have the same molecular formula but different arrangements of atoms. There are two main types of isomerism: **structural isomerism** and **stereoisomerism**.

In Unit 1 you will only be examined on structural isomerism. You will meet stereoisomerism in Unit 2.

Structural isomerism

Structural isomerism occurs when two or more molecules have the same molecular formula but different structural formulae.

There are two types of structural isomerism: **chain** and **positional**.

Chain isomerism

Chain isomerism occurs when two or more molecules have the same molecular formula but different arrangements of the carbon chain. Typically this will involve straight chain and branched chain structures.

For example there are two chain isomers of C_4H_{10} as the carbon atoms can be arranged in a straight chain forming butane or as a chain of three carbon atoms with one side chain forming methylpropane.

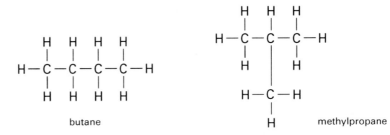

butane   methylpropane

For C_5H_{12} there are three possible chain isomers; the carbons atoms can be arranged in a straight chain forming pentane, as a chain of four atoms with a side chain forming methylbutane or as a chain of three atoms with two side chains forming dimethylpropane (the prefix di is used to show that there are two methyl groups attached to the central carbon atom).

Chain isomers of C_5H_{12}

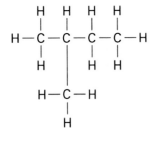

methylbutane

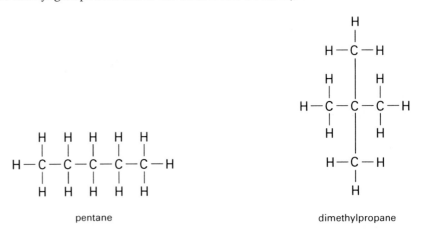

pentane   dimethylpropane

Positional isomerism

Positional isomerism occurs when a functional group can be in more than one position on the carbon chain.

You must consider positional isomerism when the molecule contains a chain of four or more carbon atoms or when the functional group is a halogen which can be in different positions on the chain.

- In butene (C_4H_8) there are two possible positions for the C=C, between atoms one and two or between atoms two and three.
- The positions are numbered so that the numbers are as small as possible.

In bromopropane (C_3H_7Br) the bromine atom can be bonded to the end carbon atom or the middle carbon atom, giving two possible positional isomers.

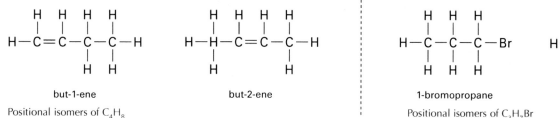

but-1-ene   but-2-ene

Positional isomers of C_4H_8

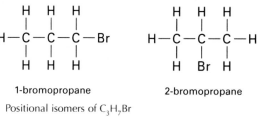

1-bromopropane   2-bromopropane

Positional isomers of C_3H_7Br

More complex examples

In an examination you are unlikely to be asked to draw a lot of isomers as it is a very time-consuming process! However, it is important that you approach drawing isomers in a systematic way to ensure that you don't miss any out. Name each one when you have drawn it to make sure that you do not draw the same molecule twice.

For C_5H_{10} there are five possible isomers; note that these are a combination of positional and chain isomers.

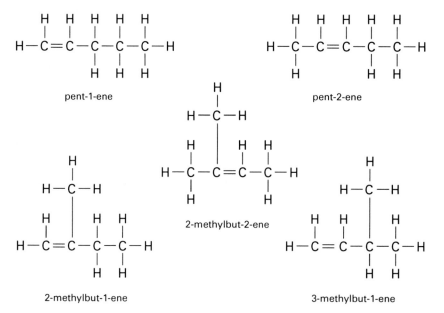

pent-1-ene   pent-2-ene

2-methylbut-2-ene

2-methylbut-1-ene   3-methylbut-1-ene

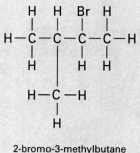
Determining molecular formula

The molecular formula of a compound is a whole number multiple of its empirical formula. The empirical formula can be calculated from percentage composition data and the molecular mass determined from a mass spectroscopy experiment.

An organic compound is analysed and found to have the following percentage composition by mass: carbon – 82.76%, hydrogen – 17.24%. The mass spectrum shows a molecular ion peak at a mass/charge value of 58.

element	C	H
% by mass	82.76	17.24
$\div A_r$	6.897	17.24
$\div$ smallest	1	2.5
$\times 2$	2	5

Empirical formula C_2H_5

The empirical formula is calculated in the same way as for an inorganic compound. Note that the final step involves multiplying the data by two in order to obtain a whole number ratio.

The mass of the empirical formula is calculated $(2 \times 12) + (5 \times 1) = 29$

The molecular mass is then divided by the empirical mass $58/29 = 2$

This tells you that the molecular formula is twice the empirical formula.

The molecular formula is C_4H_{10}.

Questions

1 Draw displayed formulae and name all the chain isomers of C_6H_{14}.

2 Draw displayed formulae and name the positional isomers of C_4H_9I.

3 Draw displayed formulae and name all the isomers of C_6H_{12}. Remember to consider both chain and positional isomers.

Hydrocarbons

Hydrocarbons are molecules containing carbon and hydrogen only.

There are several families of hydrocarbons, including alkanes, alkenes, and alkynes.

In Unit 1 you need to know some basic information about the alkane family. The alkanes can be defined as:

- saturated hydrocarbons (they contain C–C single bonds)
- molecules with the general formula C_nH_{2n+2}

Straight chain and branched chain alkanes

For alkanes with more than three carbon atoms it is possible to draw the molecules as straight or as branched chains:

- These are called chain isomers.
- Their chemistry will be very similar.
- They have slight differences in physical properties such as volatility (the ease with which a substance evaporates), melting point, and boiling point.
- These properties also vary with increasing chain length.

Look at the graph showing number of carbon atoms (i.e. chain length) and boiling point.

- As the carbon chain gets longer the boiling point increases.
- The difference between each molecule is one $-CH_2-$ unit.
  - As a result the trend is almost linear.

Explanation for this trend:

- Remember that melting point and boiling point trends are related to the forces between the molecules.
- For the alkanes (which are non-polar molecules) the intermolecular force is a van der Waals' force.
- The strength of this force increases with increasing chain length as the molecules are bigger.
  - As a result the boiling points increase.

Now look at the data (below) showing the boiling points of the isomers of C_5H_{12}:

- Note that the boiling point decreases with increased chain branching.
- This is because the surface area of the molecule is decreasing.
  - As a result the van der Waals' force is weaker.

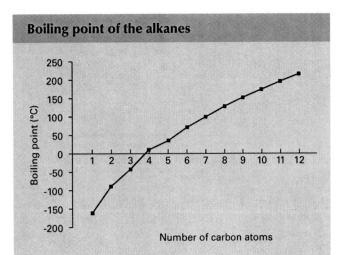

Boiling point of the alkanes

Boiling point (°C) vs Number of carbon atoms

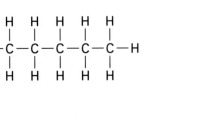

pentane
C_5H_{12}
boiling point = 36°C

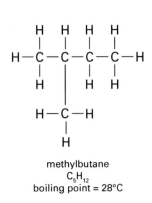

methylbutane
C_5H_{12}
boiling point = 28°C

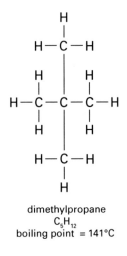

dimethylpropane
C_5H_{12}
boiling point = 141°C

Fractional distillation and the petrochemical industry

Petroleum is a mixture of hydrocarbons, most of them alkanes.

- These alkanes are separated into groups with similar boiling points.
- These groups are called **fractions**.
- The process of separating the petroleum is called **fractional distillation**.
- This relies on the differences in boiling points.

The fractionating column

You will not be asked to draw a fractionating column but it is important that you understand how it works.

- The petroleum is heated to around 350°C so that it forms a vapour.
- The vapour is then passed into the fractionating column.
- Note that the temperature of the column falls as it is ascended (this is called a negative temperature gradient).
- The largest hydrocarbons have the highest boiling points so remain as liquids at the bottom of the column.
- These are tapped off as residue.
- The smallest hydrocarbons have the lowest boiling points so rise up through the column and leave the top as gases.
- The remaining hydrocarbons rise up the column until they reach the region that corresponds to their boiling point.
- They then condense at the bubble caps and are removed from the column as liquids.

Questions

1 Think about the alkanes with molecular formula C_6H_{14}.
   a Write full displayed formulae for all the possible isomers.
   b Name each one.
   c Arrange the isomers in order of increasing boiling point starting with the lowest.
   d Make sure you can explain the order you have chosen (look back at the page on intermolecular forces if you need to).
2 Write out the name and one use for each of the fractions of petroleum. Devise your own rhyme to help you remember these, then learn the rhyme.
3 Write a three sentence explanation of how the fractions are obtained from petroleum.

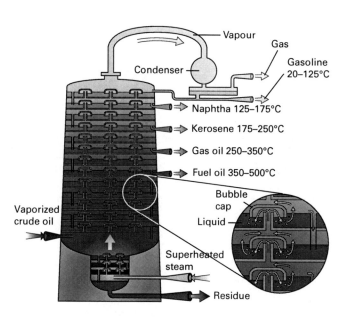

Uses of the fractions of petroleum

fraction	number of carbon atoms	typical range of boiling points (°C)	typical use of fraction
gases	1–4	below 20	camping gas
gasoline (petrol)	4–12	20–125	fuel for cars
naphtha	7–14	125–175	chemical feedstock
kerosene (paraffin)	11–15	175–250	jet fuel and chemical feedstock
gas oil (diesel)	15–19	250–350	fuel for vehicles and heating
mineral oil	*20–30*	*over 350*	*lubricating oil*
fuel oil	30–40	350–500	fuel for power stations and ships
residue: bitumen/tar			

1.18 Cracking

OBJECTIVES

By the end of this section you should be able to:

○ *write balanced equations for cracking reactions*

○ *recall the conditions for thermal cracking and catalytic cracking*

○ *understand the economic reasons for cracking*

Hexane, cyclohexane, and benzene

hexane

cyclohexane

benzene

HSW: Petrol

Petrol should ignite in an engine because of a spark at just the right time. If it ignites too soon a knocking sound which reduces efficiency of combustion is record.

The higher the octane rating of the petrol the less it is likely to cause knocking.

Why cracking?

The **fractional distillation** of petroleum produces a range of fractions with different boiling points. The mixture of fractions obtained often contains a higher proportion of heavier fractions than is needed. The **petrochemical industry** carries out the processing of these heavier fractions in order to make more useful ones.

- **Cracking** is a decomposition reaction which involves the breaking of C–C single bonds.
- This results in the formation of shorter chain alkanes and alkenes.

The heptane molecule can be cracked in a number of ways; in each equation note that both an alkane and an alkene are formed.

$$C_7H_{16} \rightarrow C_5H_{12} + C_2H_4$$
$$C_7H_{16} \rightarrow C_4H_{10} + C_3H_6$$
$$C_7H_{16} \rightarrow C_3H_8 + 2C_2H_4$$
$$C_7H_{16} \rightarrow C_5H_{10} + C_2H_4 + H_2$$

The shorter chain alkanes are more useful as fuels than the original alkane as they are more **volatile** (evaporate more readily) and burn more easily. The alkenes are used as raw materials for making polymers.

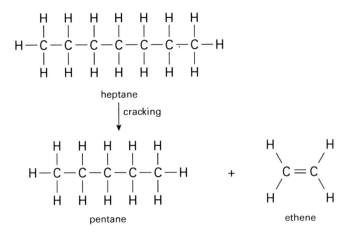

heptane

cracking

pentane

ethene

Thermal and catalytic cracking

There are two ways cracking can be carried out:

- **thermal** cracking
- **catalytic** cracking

You need to be aware of the differences in reaction conditions used to carry out the two types of cracking.

	thermal cracking	catalytic cracking
raw material	long chain alkane	long chain alkane
temperature	800–900°C	500°C
pressure	up to 7000 kPa	slightly above atmospheric pressure
catalyst	none	silica and aluminium oxide or zeolite
products	alkene, short chain alkane	aromatic hydrocarbons
uses of products	making polymers	motor fuels (short chains burn more readily, more volatile)
notes	• air is excluded from this process • heating carried out for <1 second to prevent total decomposition.	• more efficient than thermal cracking • produces more branched, cyclic, and aromatic hydrocarbons

Economic reasons for cracking

The fractional distillation process produces disproportionate amounts of some of the fractions of petroleum.

- The amount of petrol produced is not enough for our needs.
- Too much of the naphtha fraction is produced.

The cracking process helps us make maximum use of petroleum by converting the less useful fractions into more useful ones. This is an essential process, as petroleum is a **non-renewable** resource.

Petrol

Petrol is a mixture of hydrocarbons:

- Petrol manufacturers mix hydrocarbons to obtain the petrol with the right properties.
- The petrol must vaporize readily enough to ignite in the engine.
- It must not be so volatile that excess vapour enters the engine.
- It must not ignite in the cylinders of the engine too early.

These properties are temperature dependent.

- For example, longer chains are not so volatile so are used when the external temperature is warmer.
- Winter blends would vaporize too readily in the summer.

You may have heard of the term **octane rating**. This tells us how easily the petrol ignites. A lower octane rating indicates that a fuel will more readily auto-ignite. This means that the petrol ignites in the engine before it should. This causes a knocking sound and can lead to engine damage.

Processing of hexane forms three different molecules, each of which has a unique octane rating:

You do not need to remember these values but you do need to have an understanding of the difference between a low and a high octane rating.

Questions

1. Write equations using displayed formulae for the cracking of:
   a hexane, C_6H_{14}
   b octane, C_8H_{18}
2. Work through the table comparing the conditions used for thermal and catalytic cracking. Make sure you can explain why these conditions are chosen.
3. Explain in your own words the following terms:
   a octane rating
   b auto-ignition
   c volatility

name of molecule	octane rating
hexane	19
3-methylpentane	86
cyclohexane	83
benzene	106

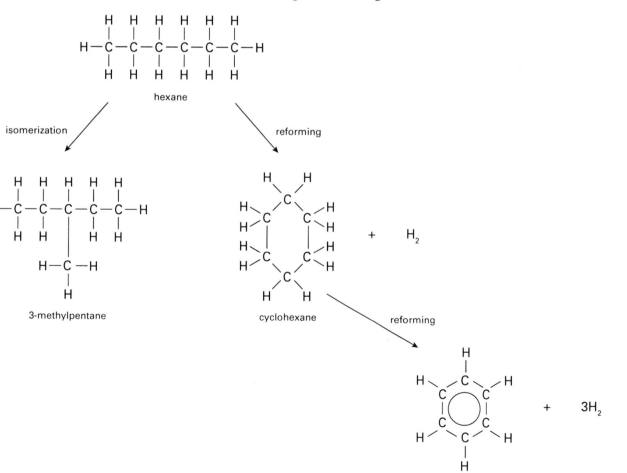

hexane

isomerization

reforming

3-methylpentane

cyclohexane

$+$ H_2

reforming

benzene

$+$ $3H_2$

1.19 Combustion

OBJECTIVES

By the end of this section you should be able to:

○ *write balanced equations for complete and incomplete combustion*

○ *describe and explain the reactions occurring inside an engine*

○ *explain how a catalytic converter works*

○ *distinguish between the environmental problems of acid rain and global warming*

○ *explain how flue-gas desulfurization works*

The catalytic converter

You will already know that a catalyst is a substance that changes the rate of a reaction but is chemically unaltered at the end of the reaction.

A catalytic converter is fitted to the exhaust system of a car.

- It removes pollutants from the exhaust gases before they are released through the end of the exhaust.
- The converter contains a honeycomb mesh of two catalysts (platinum-rhodium and platinum-palladium).
- Using a honeycomb mesh provides a large surface area for the gases to pass over.
- The catalytic converter can convert more than 90% of the pollutant gases into less harmful gases.

At the platinum-rhodium catalyst the NO_x is reduced to nitrogen.

$$2NO(g) \rightarrow N_2(g) + O_2(g)$$

At the platinum-palladium catalyst two oxidation reactions occur. Any carbon monoxide formed is oxidized to carbon dioxide.

$$CO(g) + \tfrac{1}{2}O_2(g) \rightarrow CO_2(g)$$

Unburned hydrocarbons are oxidized to carbon dioxide and water.

For example:
$$C_8H_{18}(g) + 12\tfrac{1}{2}O_2(g) \rightarrow$$
$$8CO_2(g) + 9H_2O(l)$$

Complete and incomplete combustion

The combustion of fossil fuels is occurring all the time: frying an egg, driving a car, using a computer.

Complete combustion

For **complete combustion** to occur:
- An unlimited supply of oxygen is needed.
- Alkane + oxygen → carbon dioxide + water + release of energy

For methane (the main fossil fuel in natural gas):

$$CH_4(g) + 2O_2(g) \rightarrow CO_2(g) + 2H_2O(l)$$

For octane, a longer chain hydrocarbon, the products are identical.

$$C_8H_{18}(g) + 12\tfrac{1}{2}O_2(g) \rightarrow 8CO_2(g) + 9H_2O(l)$$

Take care when balancing equations like this. Often a *half* is used to balance the equation.

Incomplete combustion

In a **limited supply** of oxygen, incomplete combustion occurs.
- The hydrogen in the hydrocarbon still forms water.
- The carbon only undergoes partial oxidation to form carbon monoxide.
- In some cases unburnt carbon particles are released as soot during incomplete combustion.
- Carbon monoxide is a toxic gas that binds to the haemoglobin in red blood cells and prevents them carrying oxygen.
- Carbon monoxide is difficult to detect because it is colourless, odourless, and tasteless.

The incomplete combustion of the same fuels as shown above gives:

$$CH_4(g) + 1\tfrac{1}{2}O_2(g) \rightarrow CO(g) + 2H_2O(l)$$
$$C_8H_{18}(g) + 8\tfrac{1}{2}O_2(g) \rightarrow 8CO(g) + 9H_2O(l)$$

Inside an engine

The combustion temperature of fuel in an engine can be 1000 °C.
- Under these conditions enough energy is available for the nitrogen and the oxygen in air to react together forming a number of oxides of nitrogen.
- These are referred to in general as NO_x.

One of the gases formed is nitrogen monoxide which can undergo further **oxidation** in the air to nitrogen dioxide.

$$N_2(g) + O_2(g) \rightarrow 2NO(g)$$
$$2NO(g) + O_2(g) \rightarrow 2NO_2(g)$$

Some of the fuel passes through the car engine without undergoing oxidation. These gases are called unburned hydrocarbons or volatile organic compounds. In sunlight these unburned hydrocarbons compounds can react with NO_x to form photochemical smog.

Global warming

Our atmosphere is made up of several gases, a number of which play an important role in controlling our climate.
- Some of these gases are described as **greenhouse gases** eg water vapour, carbon dioxide, and methane.
- These are good at absorbing infrared radiation and re-emitting in a different direction which warms the atmosphere.
- This effect is called the greenhouse effect.

The world needs a greenhouse effect to keep the planet warm enough to sustain life.

However, processes such as the burning of fossil fuels have released large amounts of greenhouse gases.

As a result our planet is getting warmer. This is **global warming** and can lead to climate change.

The greenhouse effect

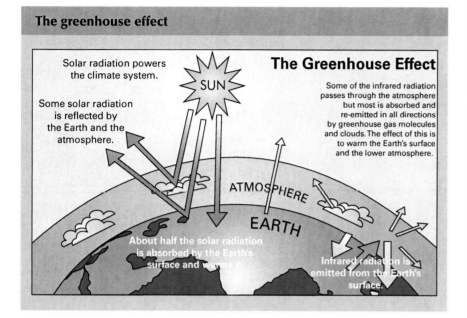

The Greenhouse Effect

Solar radiation powers the climate system.

Some solar radiation is reflected by the Earth and the atmosphere.

Some of the infrared radiation passes through the atmosphere but most is absorbed and re-emitted in all directions by greenhouse gas molecules and clouds. The effect of this is to warm the Earth's surface and the lower atmosphere.

SUN

ATMOSPHERE

EARTH

About half the solar radiation is absorbed by the Earth's surface and warms it.

Infrared radiation is emitted from the Earth's surface.

Acid rain

Our rain is naturally acidic because it contains dissolved carbon dioxide from the atmosphere. We use the term **acid rain** for rain that is more acidic than normal.

- Acid rain may be formed naturally or by human activity.
- Nitrogen dioxide formed in engines forms nitric acid:

$$2NO_2(g) + H_2O(l) + \tfrac{1}{2}O_2(g) \rightarrow 2HNO_3(aq)$$

- Fossil fuels contain sulfur which forms sulfur dioxide during combustion.
- This then dissolves in water in a reversible reaction forming sulfurous acid

$$SO_2(g) + H_2O(l) \rightleftharpoons H_2SO_3(aq)$$

- A series of reactions (which you don't need to learn) then lead to the formation of sulfuric acid, $H_2SO_4(aq)$.

Acid rain can cause in chemical weathering and acidification.

Flue-gas desulfurization

Flue gas is the waste gas from boilers and furnaces.

Flue-gas desulfurization is the process that removes sulfur dioxide from waste gases. It is used in coal-fired power stations, which release large volumes of sulfur dioxide.

A number of reactions occur during the flue-gas desulfurization process. The key equation is that for the reaction of calcium oxide, CaO, with sulfur dioxide gas forming calcium sulfite:

$$CaO(s) + SO_2(g) \rightarrow CaSO_3(s)$$

The calcium oxide is produced by heating limestone, calcium carbonate, $CaCO_3$.

$$CaCO_3(s) \rightarrow CaO(s) + CO_2(g)$$

Note that this is a far from ideal solution as the production of calcium oxide involves the release of carbon dioxide.

The calcium sulfite formed is almost insoluble in water and presents a disposal problem. It can however, be oxidized to make hydrated calcium sulfate which is gypsum, used to make plasterboard:

$$CaSO_3(s) + \tfrac{1}{2}O_2(g) + 2H_2O(l) \rightarrow CaSO_4 \cdot 2H_2O(s)$$

HSW: Climate change

*Note that it has taken many years for scientists to reach agreement about global warming and **climate change**. The Intergovernmental Panel on Climate Change has concluded that the rise in average temperatures is very likely to be due to increased greenhouse gases.*

Questions

1 Camping gas is a mixture of the alkanes propane and butane.
   a Write balanced equations for the complete combustion of butane and propane.
   b Write balanced equations for the incomplete combustion of butane and propane.
2 Write a short description of the following terms:
   a acid rain
   b the greenhouse effect
3 Practise writing out the equations for all the processes discussed in this section.

Flue gas desulfurization

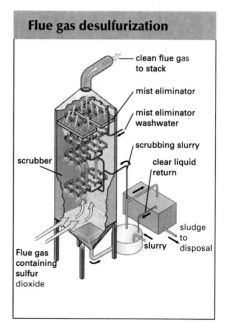

clean flue gas to stack

mist eliminator

mist eliminator washwater

scrubbing slurry

clear liquid return

scrubber

sludge to disposal

slurry

Flue gas containing sulfur dioxide

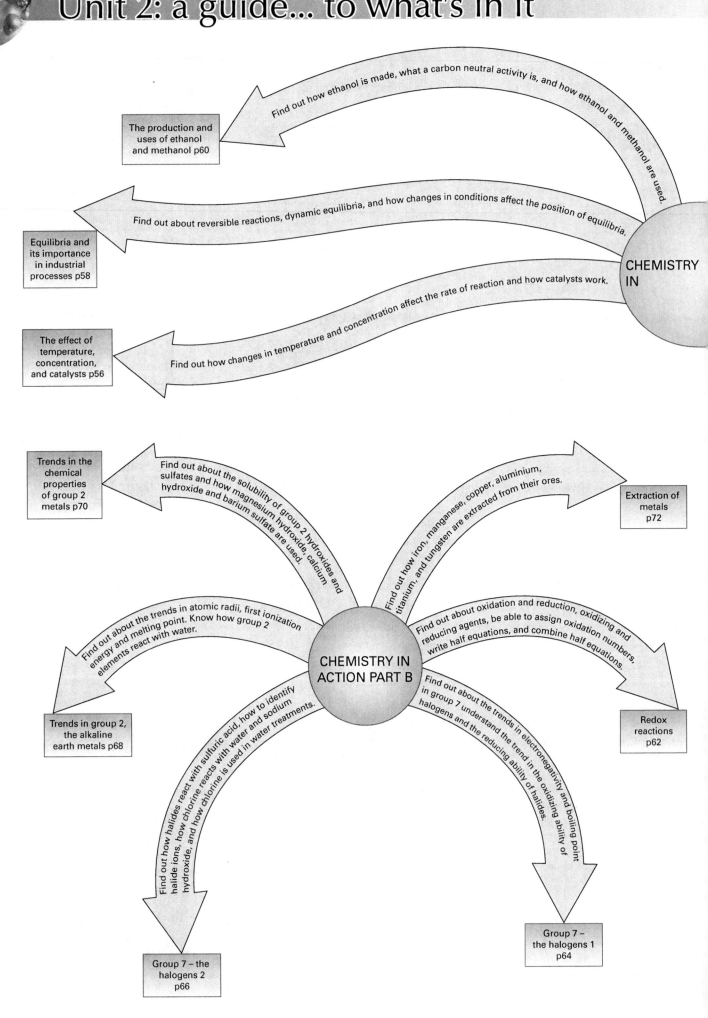

The production and uses of ethanol and methanol p60

Find out how ethanol is made, what a carbon neutral activity is, and how ethanol and methanol are used.

Equilibria and its importance in industrial processes p58

Find out about reversible reactions, dynamic equilibria, and how changes in conditions affect the position of equilibria.

CHEMISTRY IN

The effect of temperature, concentration, and catalysts p56

Find out how changes in temperature and concentration affect the rate of reaction and how catalysts work.

Trends in the chemical properties of group 2 metals p70

Find out about the solubility of group 2 hydroxides and sulfates and how magnesium hydroxide, calcium hydroxide and barium sulfate are used.

Find out how iron, manganese, copper, aluminium, titanium, and tungsten are extracted from their ores.

Extraction of metals p72

Find out about the trends in atomic radii, first ionization energy and melting point. Know how group 2 elements react with water.

Find out about oxidation and reduction, oxidizing and reducing agents, be able to assign oxidation numbers, write half equations, and combine half equations.

CHEMISTRY IN ACTION PART B

Trends in group 2, the alkaline earth metals p68

Redox reactions p62

Find out how halides react with sulfuric acid, how to identify halide ions, how chlorine reacts with water and sodium hydroxide, and how chlorine is used in water treatments.

Find out about the trends in electronegativity and boiling point in group 7 understand the trend in the oxidizing and reducing ability of halides.

Group 7 – the halogens 2 p66

Group 7 – the halogens 1 p64

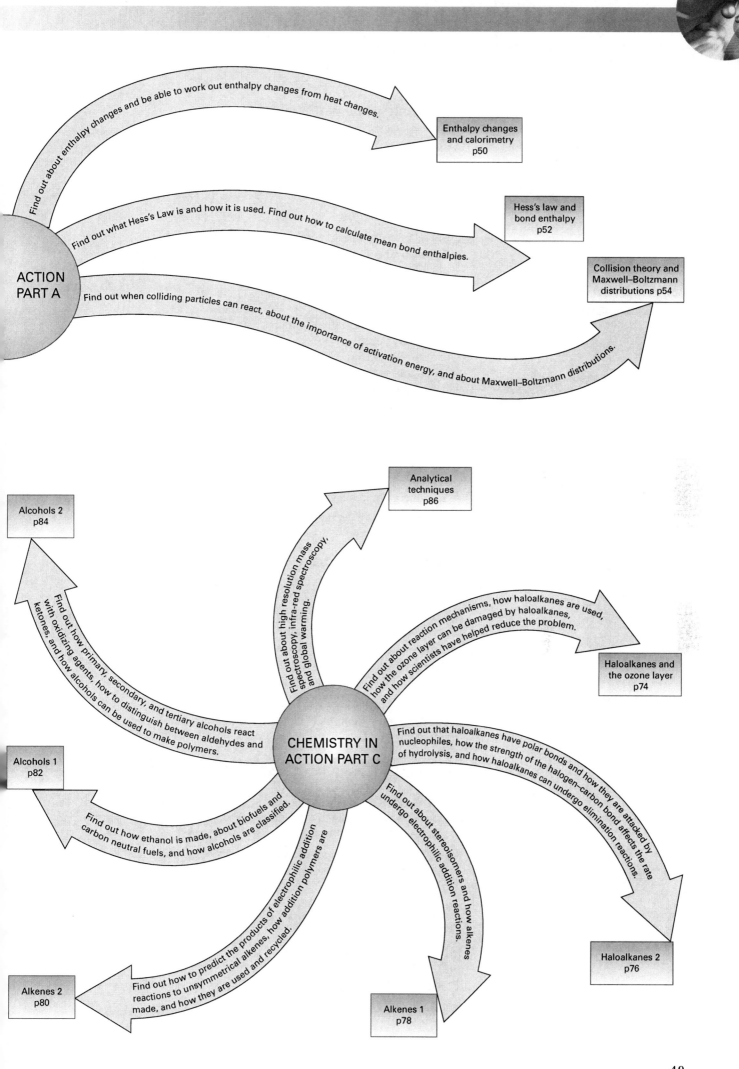

ACTION PART A

Find out about enthalpy changes and be able to work out enthalpy changes from heat changes.

Enthalpy changes and calorimetry p50

Find out what Hess's Law is and how it is used. Find out how to calculate mean bond enthalpies.

Hess's law and bond enthalpy p52

Find out when colliding particles can react, about the importance of activation energy, and about Maxwell–Boltzmann distributions.

Collision theory and Maxwell–Boltzmann distributions p54

CHEMISTRY IN ACTION PART C

Alcohols 2 p84

Find out how primary, secondary, and tertiary alcohols react with oxidizing agents, how to distinguish between aldehydes and ketones, and how alcohols can be used to make polymers.

Analytical techniques p86

Find out about high resolution mass spectroscopy, infra-red spectroscopy, and global warming.

Find out about reaction mechanisms, how haloalkanes are used, how the ozone layer can be damaged by haloalkanes, and how scientists have helped reduce the problem.

Haloalkanes and the ozone layer p74

Alcohols 1 p82

Find out that haloalkanes have polar bonds and how they are attacked by nucleophiles, how the strength of the halogen–carbon bond affects the rate of hydrolysis, and how haloalkanes can undergo elimination reactions.

Find out how ethanol is made, about biofuels and carbon neutral fuels, and how alcohols are classified.

Find out about stereoisomers and how alkenes undergo electrophilic addition reactions.

Haloalkanes 2 p76

Find out how to predict the products of electrophilic addition reactions to unsymmetrical alkenes, how addition polymers are made, and how they are used and recycled.

Alkenes 2 p80

Alkenes 1 p78

Enthalpy

- Enthalpy, H, is the heat energy stored in a chemical system.
- Enthalpy change, ΔH, is the heat energy change at constant pressure.

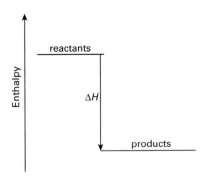

The enthalpy diagram for an exothermic reaction

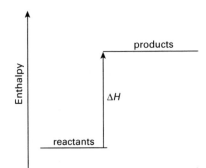

The enthalpy diagram for an endothermic reaction

Exothermic and endothermic reactions

Energy must be taken in to break existing bonds, while energy is given out when new bonds are formed. During chemical reactions there is often a difference between the amount of energy taken in to break the existing bonds and the amount of energy given out when the new bonds are formed. Most reactions give out energy overall. These reactions are described as **exothermic**.

Exothermic reactions

Burning and rusting are examples of exothermic reactions.

- The substances involved in exothermic reactions get hotter because chemical energy is being changed into thermal (heat) energy.
- During the reaction the chemicals lose energy. The energy that is lost by the chemicals is gained by the surroundings.

  ⊛ As a result the enthalpy change (ΔH) for an exothermic reaction is always negative.

The more exothermic a reaction is the more likely it is to happen, although other factors such as the activation energy will determine whether the reaction will actually take place.

Endothermic reactions

Some reactions take in more energy to break existing bonds then they release when new bonds are formed. These reactions are described as **endothermic**.

- The substances involved in endothermic reactions get colder because thermal energy is taken from the surroundings.
- The energy that is lost by the surroundings is gained by the chemicals.

  ⊛ As a result the enthalpy change (ΔH) for an endothermic reaction is always positive.

Standard enthalpy changes

The amount of energy given out or taken in depends on the temperature and pressure.

We use standard enthalpy changes so that different enthalpy changes can be compared. This means that all the other **conditions** must be kept constant.

- 1 mole of a substance is reacted.
- Any solutions have a concentration of 1.00 mol dm^{-3}.
- Any gases must have a pressure of 100 kPa.
- A stated temperature must be used, normally $25°C$.
- Elements must be in their standard states, for example carbon as graphite not as diamond.

Standard enthalpy change of formation

- The **standard enthalpy of formation**, $\Delta H_f^\ominus$ is the enthalpy change when one mole of a compound is formed from its elements in their standard states, under standard conditions.

  ⊛ As a result the standard enthalpy change of formation of an element in its standard state is zero

Standard enthalpy change of combustion

The **standard enthalpy of combustion**, $\Delta H_c^\ominus$ is the enthalpy change when one mole of a compound is completely burnt in oxygen under standard conditions.

The reaction involving $\Delta H_c^\ominus$ of methane, CH_4, is represented as

$$CH_4(g) + 2O_2(g) \rightarrow CO_2(g) + 2H_2O(g)$$

$$\Delta H_c^\ominus \text{ at } 25°C = -890.3 \text{ kJ mol}^{-1}$$

Calorimetry

Calculating the heat change of a reaction

The heat energy change for a reaction (J, measured in joules) is given by the equation

$$J = m c \Delta T$$

Where m = the mass of the surroundings (g), c = the specific heat capacity of the surroundings, ($J g^{-1} °C^{-1}$) and ΔT is the temperature change (final temperature – initial temperature), (°C).

Measuring the change in temperature

A series of temperature readings should be taken before, during, and after the reaction, using a table such as the one shown below.

Time (mins)	0.0	0.5	1.0	1.5	2.0	2.5	3.0	3.5	4.0
Temperature (°C)				X					

After 1.5 minutes the zinc was added to the copper sulfate. Notice that an X is placed on the table.

These results can be used to draw a graph. By drawing lines of best fit you can calculate the change in temperature.

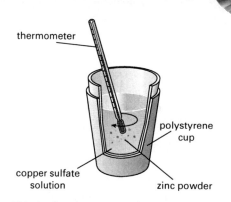

This simple calorimeter can be used to calculate the heat energy change during the displacement reaction between zinc and copper sulfate.

HSW: Calorimetry

Calorimetry can be used to calculate the heat energy change of a reaction.

Calculating the enthalpy change of a reaction

- Enthalpy changes are measured in kJ mol^{-1}

- Divide the heat change for the reaction by 1000

- Then divide the heat energy change by the number of moles involved.

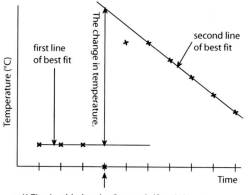

Measuring the change in temperature.

Questions

1 What happens to the thermal energy lost by the chemicals in an exothermic reaction?

2 Why do endothermic reactions get colder?

3 What is $\Delta H_f^{\ominus}$ of $N_2(g)$?

2.02 Hess's law and bond enthalpy

First law of thermodynamics

Energy cannot be created or destroyed; it is merely changed from one form to another.

Hess's law

The enthalpy change of a chemical reaction is **independent** of the route by which the reaction is achieved and depends only on the initial and final states.

- ⊕ As a result you can use **Hess's law** to find enthalpy changes for reactions that cannot be measured directly in the laboratory.
- The enthalpy change from the reactants A to the products B is the same as the sum of the enthalpy changes from A to C and from C to B.
- We can use $\Delta H_f^{\ominus}$ and $\Delta H_c^{\ominus}$.

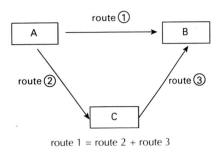

route 1 = route 2 + route 3

Using enthalpies of combustion

The standard enthalpy of combustion, ($\Delta H_c^{\ominus}$) is the enthalpy change when one mole of a compound is completely burnt in oxygen under standard conditions.

- Notice that the arrow is drawn from the chemicals to the combustion products.

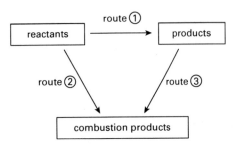

Using enthalpy of combustion values: route 1 = route 2 − route 3

Worked example

Calculate the enthalpy change for

$$C(s) + 2H_2(g) \rightarrow CH_4(g)$$

Given that

$\Delta H_c^{\ominus}$ C(s)	$= -394$ kJ mol^{-1}
$\Delta H_c^{\ominus}$ H$_2$(g)	$= -286$ kJ mol^{-1}
$\Delta H_c^{\ominus}$ CH$_4$(g)	$= -890$ kJ mol^{-1}

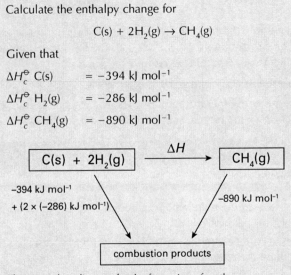

The Hess's law diagram for the formation of methane.

The unknown enthalpy change $\Delta H^{\ominus}$

$= -394$ kJ mol^{-1} + (2×-286 kJ mol^{-1}) − (-890 kJ mol^{-1})

$= -76$ kJ mol^{-1}

Worked example

Calculate the enthalpy change for the hydrogenation of ethene to form ethane.

$$C_2H_4(g) + H_2(g) \rightarrow C_2H_6(g)$$

Given that

$\Delta H_c^{\ominus}$ C$_2$H$_4$(g)	$= -1409$ kJ mol^{-1}
$\Delta H_c^{\ominus}$ H$_2$(g)	$= -286$ kJ mol^{-1}
$\Delta H_c^{\ominus}$ C$_2$H$_6$(g)	$= -1560$ kJ mol^{-1}

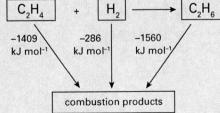

The Hess's law diagram for the hydrogenation of ethene.

The unknown enthalpy change $\Delta H^{\ominus}$

$= -1409$ kJ mol^{-1} -286 kJ mol^{-1} − (-1560 kJ mol^{-1})

$= -135$ kJ mol^{-1}

Using enthalpies of formation

The standard enthalpy of formation, $\Delta H_f^{\ominus}$ is the enthalpy change when one mole of a compound is formed from its elements in their standard states under standard conditions.

- Notice that the arrow is drawn from the elements to the chemicals.

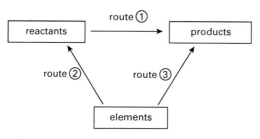

Using standard enthalpy of formation values:
route 1 = − route 2 + route 3

Worked example

Calculate the enthalpy change for the combustion of ethene, $C_2H_4(g)$.

$$C_2H_4(g) + 3O_2(g) \rightarrow 2CO_2(g) + 2H_2O(g)$$

Given that

$\Delta H_f^{\ominus} C_2H_4(g) = +52 \text{ kJ mol}^{-1}$

$\Delta H_f^{\ominus} CO_2(g) = -394 \text{ kJ mol}^{-1}$

$\Delta H_f^{\ominus} H_2O(g) = -286 \text{ kJ mol}^{-1}$

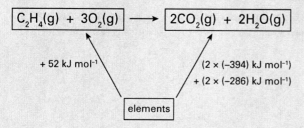

The Hess's law diagram for the combustion of ethene.

The unknown enthalpy change $\Delta H^{\ominus}$

$= - (+ 52 \text{ kJ mol}^{-1}) + (2 \times -394 \text{ kJ mol}^{-1})$
$\quad + (2 \times - 286 \text{ kJ mol}^{-1})$

$= -1412 \text{ kJ mol}^{-1}$

Mean bond enthalpies

The **mean (average) bond enthalpy** is the mean amount of energy required to break one mole of a specified type of covalent bond in a gaseous species.

$$A-B(g) \rightarrow A(g) + B(g)$$

- This average is obtained from the values obtained from many molecules so the actual values will probably be a little different.
- Energy is required to break bonds.
- Different bonds may require different amounts of energy

$$CH_4(g) \rightarrow C(g) + 4H(g)$$
$$\Delta H = + 1664 \text{ kJ mol}^{-1}$$

The average bond enthalpy for a C–H bond

$$= \frac{1664 \text{ kJ mol}^{-1}}{4} = 416 \text{ kJ mol}^{-1}$$

- Bond enthalpies are always endothermic.
- Energy is released when new bonds are formed.

Worked example

Calculate the enthalpy change when methane, $CH_4(g)$ is burnt.

bond	average bond enthalpy (kJ mol^{-1})
C–H	413
O=O	498
C=O	805
O–H	464

$$CH_4(g) + 2O_2(g) \rightarrow CO_2(g) + 2H_2O(g)$$

Energy required to break the existing bonds

$4 \times$ C–H $= 4 \times 413$ kJ mol^{-1} $=1652$ kJ mol^{-1}

$2 \times$ O=O $= 2 \times 498$ kJ mol^{-1} $= 996$ kJ mol^{-1}

Total $= 2648$ kJ mol^{-1}

Energy released when the new bonds are formed

$2 \times$ C=O $= 2 \times 805$ kJ mol^{-1} $=1610$ kJ mol^{-1}

$4 \times$ O–H $= 4 \times 464$ kJ mol^{-1} $= 1856$ kJ mol^{-1}

Total $= 3466$ kJ mol^{-1}

The net energy change $= 2648$ kJ mol^{-1} $- 3466$ kJ mol^{-1}
$= -818$ kJ mol^{-1}

Questions

1 What is Hess's law?

2 Why do we use Hess's law?

3 What is mean bond enthalpy?

OBJECTIVES

By the end of this section you should be able to:

○ understand that reactions can only occur when collisions take place and that these collisions must have enough energy to react

○ recall the definitions for activation energy and understand what activation means.

○ understand Maxwell–Boltzmann distributions of molecular energies of gases

Collision theory

You can use the **collision theory** to understand how the conditions used affect the rate of a chemical reaction.

• For a reaction to occur particles must collide.

Particles must collide before they can react.

• When the particles collide they must have enough energy to break the existing bonds.

• When the particles collide they must collide in the correct **orientation** so that the reactive parts of the molecules come together. This is particularly important for large molecules.

　⊛ As a result only a small proportion of collisions between particles result in a reaction.

Activation energy

Activation energy, E_a, is the minimum collision energy that particles must have to react.

• E_a is the minimum amount of energy that the reactants must have to form an **activated complex** in a transition state. In an activated complex the old bonds are partially broken and the new bonds are partially made.

• Different reactions have different activation energies.

• The lower the activation energy the larger the number of particles that can react at any temperature.

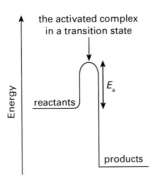

The energy change in an exothermic reaction

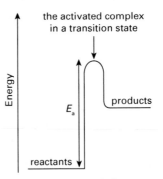

The energy change in an endothermic reaction

The effect of changing the temperature

• As the temperature of a sample of particles changes the kinetic energy of the particles changes.

• As you increase the temperature of a sample you increase the kinetic energy of the particles.

• As the temperature increases, the particles move faster so they collide more often.

• Also when the particles do collide more of the particles will have enough energy to react.

　⊛ As a result the higher the temperature the larger the number of particles that can react.

However, the particles will only react if they collide in the correct orientation.

Maxwell–Boltzmann distributions

These diagrams are used to represent the energy of the particles in a sample of a gas at a given temperature.

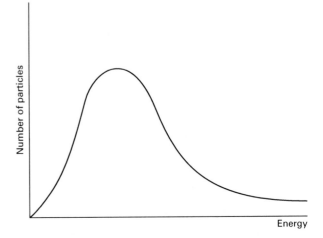

The energy distribution curve for a sample of gas at a particular temperature.

• Notice that the **distribution curve** is not symmetrical.

• Most of the particles have an energy which falls within quite a narrow range, with few particles having much more or much less energy.

• The line does not cross the x-axis at higher energy. In fact it would only do so at infinity.

• The line starts at the origin, showing that none of the particles has no energy.

• The total area under the distribution curve represents the total number of gas particles.

Activation energy and Maxwell–Boltzmann distributions

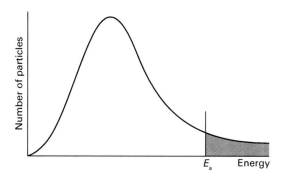

The activation energy is the minimum collision energy that particles must have to react.

- For a reaction to occur the particles must collide and these collisions must have energy equal to or greater than the activation energy. If there is not enough energy when the particles collide then the particles will not react.
- Only the small proportion of particles in the shaded part of the distribution curve have enough energy to react.
- Notice that the activation energy is drawn towards the right of the peak of the distribution curve. If more than half the particles had enough energy to react, the reaction would be too fast to be controlled safely.

Maxwell–Boltzmann distributions and temperature

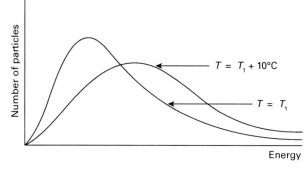

The Maxwell–Boltzmann distributions at different temperatures.

Notice how the range of energies that the particles have increases as the temperature increases.

- The average energy of the particles increases.
- The peak of the distribution curve moves to the right as the average energy increases.
- The distribution curve becomes flatter because the total number of particles remains the same.

Worked example

The graph below shows the Maxwell–Boltzmann distribution for a sample at a particular temperature T_1.

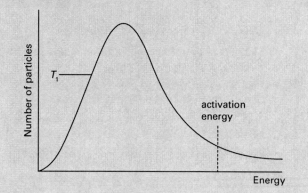

a Shade the proportion of particles that have enough energy to react at temperature T_1.

b Sketch a curve to show the Maxwell–Boltzmann distribution of energies at temperature T_2, which is 10°C hotter than T_1.

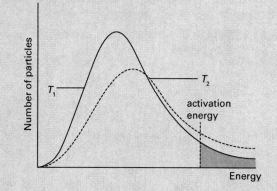

- Notice that only the particles that have energy greater than the activation energy have enough energy to react.
- Notice how at the higher temperature T_2, the average energy of the particles has increased so the peak of the graph has moved to the right and the curve has become flatter.

Questions

1 What is the activation energy of a reaction?

2 How does increasing the temperature affect the energy of the particles?

3 What do Maxwell–Boltzmann diagrams show?

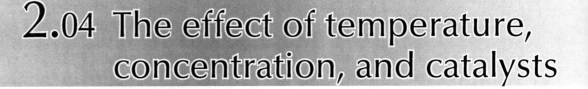

OBJECTIVES

By the end of this section you should be able to:

○ *understand the qualitative effect of temperature changes on the rate of a chemical reaction*

○ *understand how small temperature changes can lead to big increases in the rate of a chemical reaction*

○ *understand how changes in concentration affect the rate of a chemical reaction*

○ *understand what a catalyst is and how catalysts work*

The effect of changing the temperature

- For a reaction to occur the reactant particles must collide and have enough energy to react.
- As the temperature increases the particles move faster so they collide more often.
- The collisions between the particles have more energy, so more particles have enough energy to react.
  - Ⓔ As a result, as the temperature increases the rate of reaction increases.

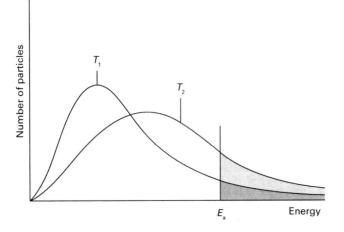

The Maxwell–Boltzmann distributions at temperature T_1 and the higher temperature T_2.

- Notice that as the temperature increases a much higher proportion of the particles have energy greater than or equal to the activation energy.
  - Ⓔ As a result even a modest increase in temperature can lead to a very large increase in the rate of a chemical reaction.

The effect of changing the concentration

- For a reaction to occur the particles must collide.
- As the concentration of a solution increases the distance between the reactant particles decreases.
- The particles have to travel less distance before colliding so they collide more often.
  - Ⓔ As a result as the concentration increases the rate of reaction increases.

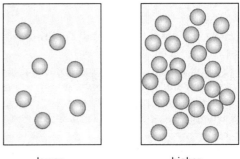

| lower concentration | higher concentration |

Increasing the concentration decreases the distance between the particles.

Catalysts

- A catalyst increases the rate of a chemical reaction but is not used up itself during the reaction.
- Catalysts work by providing an alternative reaction pathway which has **lower activation energy**.
  - Ⓔ As a result at any given temperature more particles will have energy greater than or equal to the activation energy.
  - Ⓔ As a result the reactions happen faster (the rate of reaction increases).

Catalysts and Maxwell–Boltzmann distributions

- Adding a catalyst does not change the distribution curves but it does mean that more particles have enough energy to react.

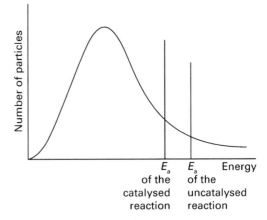

A catalysed reaction has a lower activation energy.

Catalysts and energy level diagrams

Uncatalysed reactions

- In an uncatalysed reaction the activation energy is the minimum amount of energy that must be supplied to form an activated complex in a **transition state**.

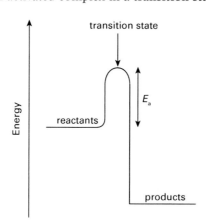

The activation energy in an uncatalysed exothermic reaction.

Catalysed reactions

- In catalysed reactions the reactants form an **intermediate**.
- The intermediate then reacts to form a product.
- The activation energy for a catalysed reaction is lower than that for an uncatalysed reaction.
  - As a result a catalyst increases the rate of a reaction.

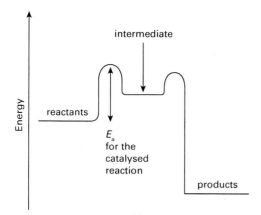

The activation energy in a catalysed exothermic reaction.

Worked example

The graph below represents the Maxwell–Boltzmann distribution of energies at a particular temperature.

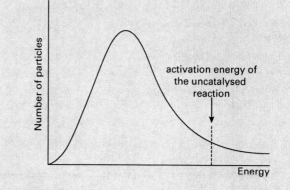

State the effect of adding a catalyst on:

a the energy of the particles

b the activation energy

c the rate of reaction

Answer

a Adding a catalyst will not affect the energy of the particles.

b Adding a catalyst will provide an alternative reaction pathway with a lower activation energy

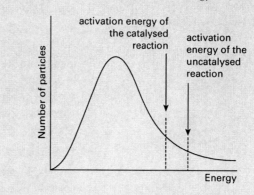

c The rate of reaction will increase because more particles now have enough energy to react.

Questions

1 Why does increasing the temperature increase the rate of a chemical reaction?

2 Why does increasing concentration increase the rate of a chemical reaction?

3 Why does adding a catalyst increase the rate of a chemical reaction?

2.05 Equilibria and its importance in industrial processes

OBJECTIVES

By the end of this section you should be able to:

○ *understand that many chemical reactions are reversible*

○ *understand what a dynamic equilibrium is*

○ *use Le Chatelier's principle*

○ *explain how catalysts and changes in temperature affect the position of equilibrium*

○ *apply these principles to chemical processes and know why compromise temperatures and pressures may be used*

Le Chatelier's principle

Le Chatelier's principle states that when the conditions of a dynamic equilibrium are changed then the position of equilibrium will shift to minimize the change.

Reversible reactions and dynamic equilibria

Many reactions are **reversible;** they can go forwards or backwards.
For example

$$SO_2(g) + \tfrac{1}{2}O_2(g) \rightleftharpoons SO_3(g)$$

- The ⇌ sign means that the reaction is reversible.
- If the forward reaction is exothermic the enthalpy change will have a negative sign. The reverse reaction will be endothermic and the enthalpy change will have a positive sign.
- Both reactants and products are constantly being made and broken up.
- If a reversible reaction takes place in a **'closed system'** where nothing can enter or leave then an **equilibrium** can be reached.
- At equilibrium there is a balance between the reactants and the products and the concentrations of the products and reactants stay the same. This does not mean that the concentration of reactants is equal to the concentration of the products.
- The forward and backward reactions have not stopped. It is just that at equilibrium the rate of forward reaction is equal to the rate of backward reaction.
  - ⊚ As a result the equilibrium is described as being dynamic.

Le Chatelier's principle

Le Chatelier's principle helps us to predict the effect of changing factors that affect the position of equilibria.

The effect of catalysts on the position of equilibrium

A catalyst is a chemical which speeds up the rate of a chemical reaction but is not itself used up. This means that catalysts can be used many times. Transition metals and transition metal compounds are often good catalysts. Catalysts are widely used in industry to reduce the costs of producing chemicals.

- Catalysts increase the rate of chemical reaction by offering an alternative reaction pathway with a lower activation energy.
- Catalysts do not affect the position of equilibrium.

The effect of changing temperature on the position of equilibrium

If a reaction is at equilibrium then changing the temperature may affect the position of equilibrium.

- The position of equilibrium will shift to minimize the change in temperature.
  - ⊚ As a result if the temperature is increased the position of equilibrium will shift in the endothermic direction.

Example

The reaction between A and B to form C is exothermic A + B ⇌ C

If the temperature is increased then the position of equilibrium will shift in the endothermic direction. At equilibrium there will be a lower yield of C.

⊚ As a result if the temperature is decreased the position of equilibrium will shift in the exothermic direction.

Example

The reaction between X and Y to form Z is endothermic X + Y ⇌ Z

If the temperature is decreased then the position of equilibrium will shift in the exothermic direction. At equilibrium there will be a lower yield of Z.

The effect of changing the pressure on the position of equilibrium

If a reaction which involves gases is at equilibrium then changing the pressure may affect the position of equilibrium.

- The position of equilibrium shifts to minimize the change.
  - If the pressure is increased it will shift the position of equilibrium towards the side which has fewer gas molecules.
  - If the pressure is decreased it will shift the position towards the side with more gas molecules.

The effect of changing the concentration on the position of equilibrium

If a reaction involving solutions is at equilibrium then changing the **concentration** will affect the position of equilibrium.

- The position of equilibrium shifts to minimize the change.
  - If the concentration of one of the reactants is increased it will shift the position of equilibrium towards the right so more product will be made.
  - If the concentration of one of the reactants is decreased it will shift the position of equilibrium towards the left so fewer products will be made.

Equilibria and chemical processes

The Haber Process

Ammonia is used to produce fertilizers and explosives.

Ammonia, NH_3, is produced by the Haber Process. The reaction is reversible and the forward reaction is exothermic.

$$N_2(g) + 3H_2(g) \rightleftharpoons 2NH_3(g)$$

- An iron catalyst is used to increase the rate of reaction. This allows us to carry out the reaction at a reasonable temperature.
- The catalyst does not affect the position of equilibrium, so it has no effect on the yield of ammonia at equilibrium.
- The forward reaction between nitrogen and hydrogen is exothermic. Increasing the temperature increases the rate of reaction but decreases the yield of ammonia (increasing the temperature shifts the position of equilibrium in the endothermic direction).
  - As a result a compromise temperature is used. This gives a reasonable rate and a reasonable yield of ammonia.
- Increasing the pressure increases the yield of ammonia (increasing the pressure shifts the position of equilibrium towards the products side, which has fewer gas particles).

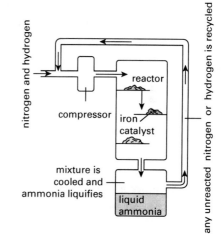

The Haber Process

The contact process

Sulfuric acid is produced by the contact process. The reaction is reversible and the forward reaction is exothermic.

$$SO_2(g) + \tfrac{1}{2}O_2(g) \rightleftharpoons SO_3(g)$$

- A vanadium(V) oxide catalyst is used to increase the rate of reaction. This allows us to carry out the reaction at a reasonable temperature.
- The catalyst does not affect the position of equilibrium, so it has no effect on the yield of sulfur trioxide at equilibrium.
- The forward reaction between sulfur dioxide and oxygen is exothermic. Increasing the temperature increases the rate of reaction but decreases the yield of sulfur trioxide (increasing the temperature shifts the position of equilibrium in the endothermic direction).
  - As a result a compromise temperature is used. This gives a reasonable rate and a reasonable yield of sulfur trioxide.
- Increasing the pressure increases the yield of sulfur trioxide (increasing the pressure shifts the position of equilibrium towards the products side, which has fewer gas particles).
- In practice a high yield is obtained without a high pressure. As maintaining a high pressure is very expensive, a lower pressure is used.

Questions

1 What does the sign $\rightleftharpoons$ mean?

2 What does Le Chatelier's principle state?

3 How does adding a catalyst affect the position of equilibrium?

By the end of this section you should be able to:

○ recall how the hydration of ethene can be used to make ethanol

○ recall that carbon monoxide can be reacted with hydrogen to make ethanol

○ understand what a 'carbon-neutral' activity is

○ recall that methanol and ethanol are useful fuels

The hydration of ethene

- **Ethanol** is a type of alcohol
- Ethanol can be made by the **hydration** of ethene. In this reaction steam is added to ethene.

Ethene + steam ⇌ ethanol vapour

$$C_2H_4(g) + H_2O(g) \rightleftharpoons C_2H_5OH(g)$$

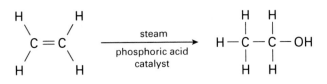

The reaction is reversible and the forward reaction is exothermic.

- A **phosphoric acid** catalyst is used. This increases the rate of reaction so a dynamic equilibrium is reached more quickly.
- The catalyst does not affect the position of equilibrium so it does not affect the yield of ethanol at equilibrium.
- The forward reaction, between ethene and steam, is exothermic.
- If the system is at equilibrium and the temperature is changed the position of equilibrium will shift to minimize the change in temperature.
- Increasing the temperature increases the rate of reaction but decreases the yield of ethanol (increasing the temperature shifts the position of equilibrium in the endothermic direction).
- Decreasing the temperature increases the yield of ethanol (decreasing the temperature shifts the position in the exothermic direction) but decreases the rate of reaction.

  ℗ As a result a compromise temperature of around 300°C is used. This gives a reasonable rate and a reasonable yield of ethanol.

If the system is at equilibrium and the pressure is changed the position of equilibrium will shift to minimize the change in pressure.

- Increasing the pressure shifts the position of equilibrium towards the side with fewer gas molecules so increases the yield of ethanol.
- Decreasing the pressure shifts the position of equilibrium towards the side with more gas molecules decreasing the yield of ethanol.
- However, it is very expensive to maintain a very high pressure. Also at very high pressures the ethene molecules may polymerize to form poly(ethene).

  ℗ As a result a pressure of 90 atmospheres is used.

The production of methanol

- Methanol is another type of alcohol.

methanol

- **Methanol** can be made by reacting carbon monoxide with hydrogen.

Carbon monoxide + hydrogen ⇌ methanol

$$CO(g) + 2H_2(g) \rightleftharpoons CH_3OH(g)$$

- The reaction is reversible and is used to store hydrogen, which can be used as a fuel, as the liquid methanol.
- If the system is at equilibrium and the concentration of one of the reactants or products is changed the position of equilibrium will shift to minimize the change in concentration.
- If an equilibrium mixture is placed in a fuel tank, as the hydrogen is used as a fuel it is removed from the mixture and the concentration of hydrogen is reduced.
- As the concentration of hydrogen is reduced the position of equilibrium shift towards the reactants side so more hydrogen is produced.

  ℗ As a result if an equilibrium mixture of carbon monoxide, hydrogen, and methanol is placed in a fuel tank, as the hydrogen is used then more hydrogen will be produced.

Carbon-neutral

Carbon dioxide is a greenhouse gas which contributes towards **global warming**.

- Large amounts of carbon dioxide are released into the atmosphere when fossil fuels such as coal, oil, and natural gas are burnt.
- Many people are trying to reduce the amount of carbon dioxide gas they are releasing into the atmosphere.
- A **carbon-neutral activity** is an activity that produces no net carbon emissions to the atmosphere over the course of a year.
- However, even in a carbon-neutral activity at some stages carbon dioxide is absorbed from the atmosphere and at other stages it is emitted back to the atmosphere.
- Overall a carbon-neutral fuel does not release carbon dioxide into the atmosphere.

- Methanol can used as a fuel for motor vehicles.
- The production of methanol from carbon monoxide and hydrogen is an example of a carbon-neutral fuel if the hydrogen is produced from a renewable source such as solar power and the carbon dioxide produced when the methanol is burnt is recycled.
- By recycling the carbon dioxide rather than releasing it into the atmosphere we are not increasing the levels of carbon dioxide gas in the environment.
- Alternatively the fuel could be made carbon-neutral by making the carbon monoxide from plant waste because the plants have absorbed carbon dioxide as they grew and when the fuel is burnt we are just releasing this carbon dioxide back into the atmosphere.

Using ethanol and methanol as fuels

The alcohols methanol and ethanol are both liquids at room temperature. Methanol and ethanol are both useful fuels. Liquid fuels are much easier to handle and store than gaseous fuels such as hydrogen and methane:

- Liquid fuels are easier to transfer than gases.
- Liquids fuels have a smaller volume than gases so smaller fuel tanks can be used.
- Also gaseous fuels would need to be kept under high pressures, so they are more likely to escape than liquids.

Ethanol made by the **fermentation** of plant material such as sugar cane is a **renewable fuel**.

- Ethanol can also be made by the hydration of ethene. As the ethene is obtained from crude oil this is non-renewable.
- Ethanol and methanol both burn very cleanly, releasing very little carbon monoxide.
- Methanol is normally produced from methane, which is obtained from natural gas so is non-renewable.
- Methanol can also be made from the reaction between hydrogen and carbon monoxide. See section *Production of methanol* (opposite page)

Worked example

Give two reasons why motorists may prefer liquid fuels such as ethanol to gaseous fuels such as hydrogen.

Answer

Liquid fuels are easier to transfer to the vehicle and because liquids take up a smaller volume than gases smaller fuel tanks can be used.

Questions

1 What is the symbol equation for the hydration of ethene?
2 What is the symbol equation for the reaction between carbon monoxide and hydrogen to make methanol?
3 What is a carbon-neutral activity?

2.07 Redox reactions

OBJECTIVES

By the end of this section you should be able to:

○ define oxidation and reduction in terms of electron loss and gain

○ recall what oxidizing agents and reducing agents are

○ assign oxidation states to the elements in a compound

○ understand the redox reactions of s and p block elements

○ write half-equations and identify the species being oxidized and reduced

○ combine half-equations to give an overall equation

Oxidation and reduction

- **Oxidation** is the loss of electrons.
- **Reduction** is the gain of electrons.
- As one species loses electrons another species must gain these electrons.
  - Ⓡ As a result oxidation and reduction must always occur together. These reactions are often called **'redox reactions'**.

Oxidizing and reducing agents

- An **oxidizing agent** is a species which oxidizes another substance by removing electrons from it.
- Oxidizing agents are themselves reduced during the reaction.
- Common oxidizing agents include oxygen, chlorine, potassium dichromate(VI), and potassium manganate(VII).
- A **reducing agent** is a species which reduces another substance by adding electrons to it.
- Reducing agents are themselves oxidized during the reaction.
- Common reducing agents include group 1 and 2 metals, hydrogen, carbon, and carbon monoxide.

Assigning oxidation states

- Any pure element has an **oxidation state** of zero.

Example

The oxidation number of nitrogen in N_2 is zero.

- Any monatomic ion has an oxidation state equal to the charge on the ion.

Example

The oxidation state of a Mg^{2+} ion is $2+$.

- In compounds, group 1 metal atoms have an oxidation state of $1+$.

Example

The oxidation state of potassium in KNO_3 is equal to $1+$.

- In compounds, group 2 metal atoms have an oxidation state of $2+$.

Example

The oxidation state of calcium in $CaCO_3$ is equal to $2+$.

- In compounds, the most electronegative element fluorine always has an oxidation state of $1-$.

Example

The oxidation state of fluorine in HF is $1-$.

- In compounds, hydrogen has an oxidation state of $1+$ unless it is in a metal hydride when it has an oxidation state of $1-$.

Examples

The oxidation state of hydrogen in H_2O is $1+$.

The oxidation state of hydrogen in KH is $1-$.

- In compounds, the oxidation state of oxygen is $2-$ unless it is with fluorine when it has an oxidation state of $2+$ or is part of peroxide when it has an oxidation state of $1-$.

Examples

The oxidation state of oxygen in H_2O is $2-$.

The oxidation state of oxygen in F_2O is $2+$.

The oxidation state of oxygen in H_2O_2 is $1-$.

- If a molecule is neutral overall, e.g. CO_2, then the sum of the oxidation states of the atoms in the molecule must be zero.
- Where a molecular ion has an overall charge, e.g. CO_3^{2-}, then the sum of the oxidation states of the atoms in the molecule must equal the overall charge of the ion.

Examples of applying oxidation states

Sodium nitrate

Sodium nitrate has the formula $NaNO_3$.

- Sodium nitrate has no overall charge so the sum of the oxidation numbers of the atoms must be equal to zero.
- There are three oxygen atoms which each have a $2-$ charge.
- Sodium is in group 1 of the period table so it has a $1+$ charge.
  - Ⓡ As a result the oxidation state of nitrogen must be $5+$.

Carbonate

A carbonate ion has the formula CO_3^{2-}.

- Carbonate has an overall charge of $2-$ so the sum of the oxidation numbers of the atoms must be equal to $2-$.
- There are three oxygen atoms which each have a $2-$ charge.
  - Ⓡ As a result the oxidation state of carbon must be $4+$.

Identifying oxidation and reduction reactions

When magnesium is added to copper sulfate solution a displacement reaction takes place.

The overall equation for the reaction is

$$Mg + CuSO_4 \rightarrow MgSO_4 + Cu$$

- The oxidation state of uncombined magnesium is zero; in $MgSO_4$ it is 2+. The oxidation state has gone up so the magnesium is oxidized.
- The oxidation state of copper is 2+ in $CuSO_4$ and zero in uncombined copper. The oxidation state has gone down so the copper has been reduced.
- We can show the electron transfer by splitting the overall equation into two half-equations.

$$Mg \rightarrow Mg^{2+} + 2e^- \text{ (oxidation)}$$

$$Cu^{2+} + 2e^- \rightarrow Cu \text{ (reduction)}$$

The magnesium is oxidized (oxidation is loss of electrons) and the copper is reduced (reduction is gain of electrons).

The magnesium has reduced the copper so it is a reducing agent.

The copper sulfate has oxidized the magnesium so it is an oxidizing agent.

Combining half-equations

We can combine half-equations to give the overall equation for the reaction.

Before we can combine the half-equations one or both of the equations may need to be multiplied by a small number so that we have equal numbers of electrons in both of the equations.

The salt iron(III) chloride is made by heating iron in chlorine gas.

- The iron is oxidized and this can be represented by the half-equation

$$Fe \rightarrow Fe^{3+} + 3e^-$$

- The chlorine is reduced and this can be represented by the half-equation

$$Cl_2 + 2e^- \rightarrow 2Cl^-$$

- Before the two half-equation can be combined we must ensure that there are equal numbers of electrons in both half-equations.

Notice that the first half-equation is multiplied by 2 to give

$$2Fe \rightarrow 2Fe^{3+} + 6e-$$

While the second half-equation is multiplied by 3 to give

$$3Cl_2 + 6e^- \rightarrow 6Cl^-$$

The two half-equations can then be added together to give the overall equation for the reaction.

$$2Fe + 3Cl_2 \rightarrow 2Fe^{3+} + 6Cl^-$$

$$\text{Or } 2Fe + 3Cl_2 \rightarrow 2FeCl_3$$

Notice that the electrons cancel out and are not written in the overall equation for the reaction.

Worked example

State the oxidation number of sulfur in

a H_2SO_4

b H_2S

c SO_2

Answer

a H_2SO_4 So S must be 6+

$1+ \times 2$ $2- \times 4$
$= 2+$ $= 8-$

b H_2S So S must be 2–

$1+ \times 2$
$= 2+$

c SO_2 So S must be 4+

$2- \times 2$
$= 4-$

State the oxidation number of iodine in

a KIO_3

b KI

c KIO_4

Answer

Questions

1 What is the oxidation state of sulfur in SO_4^{2-}?

2 What is the oxidation state of oxygen in O_2?

3 In the reaction $Fe + CuSO_4 \rightarrow FeSO_4 + Cu$ which species is oxidized and which is reduced?

OBJECTIVES

By the end of this section you should be able to:

○ *understand the trends in electronegativity and boiling point in group 7*

○ *understand the trend in oxidizing ability of the halogens*

○ *understand the trend in the reducing ability of halide ions*

Introducing the halogens

- The elements in group 7 of the periodic table are often called the **halogens**.

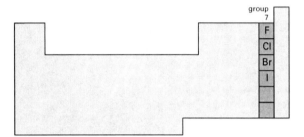

Group 7 elements are typical non-metals.

- Halogen atoms have seven electrons in their outer shell.

Example

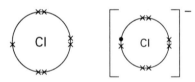

When halogen atoms react they can gain an electron to form a halide ion which has a 1− charge.

Trends in electronegativity

Electronegativity is a way of measuring the attraction that a bonded atom has for the electrons in a covalent bond.
Down the group the halogens become less electronegative.

Trends in boiling point

element	boiling point (°C)
fluorine	−188
chlorine	−35
bromine	+59
iodine	+216

At room temperature:
- Fluorine is a yellow gas.
- Chlorine is a pale green gas.
- Bromine is a brown liquid.
- Iodine is a dark grey solid.

Notice how the boiling points of the halogens increase down the group.

- The halogens exist as diatomic molecules, e.g. chlorine exists as chlorine molecules, Cl_2.
- This means that there are strong covalent bonds within the halogen molecules but only very much weaker van der Waals' forces of attraction between halogen molecules.
- The strength of the van der Waals' forces depends on the number of electrons in the molecules.
- Down the group the number of electrons in the halogen molecules increases.
  - As a result down the group the boiling point increases.

Worked example

Complete the table below

element	state at room temperature	colour
fluorine		
chlorine		
bromine		
iodine		

element	state at room temperature	colour
fluorine	gas	yellow
chlorine	gas	pale green
bromine	liquid	brown
iodine	solid	dark grey

Worked example

Explain why iodine has a higher boiling point than fluorine.

Answer

Down the group the boiling point of the elements increases because there are stronger intermolecular forces between the molecules. Weak van der Waals' forces exist between the halogen molecules. Down the group there are more electrons in the atoms so there are stronger van der Waals' forces.

More energy is required to overcome these forces of attraction so the halogen boils at a higher temperature.

Trends in the oxidizing ability of the halogens

The **oxidizing ability** of a substance is a measure of the strength of an atom to attract and gain an electron.

- The halogens are good oxidizing agents.
- An oxidizing agent is a species which oxidizes another substance by removing electrons from it.
- Oxidizing agents are themselves reduced during the reaction.

When halogen atoms react they gain electrons.

Example

$$Cl_2 + 2e^- \rightarrow 2Cl^-$$

The oxidizing ability of the halogens decreases down the group.

Down the group the electron which is gained is being placed into a shell which is further from the nucleus.

- Down the group the atomic radii increases.
- The amount of shielding increases.
  - As a result the attraction between the nucleus and the electron decreases.
  - As a result the oxidizing ability of the halogens decreases down the group.

Displacement reactions

- The **displacement** reactions between halogens and aqueous halides demonstrate the decrease in the oxidizing ability of the halogens down the group.
- Chlorine is a more powerful oxidizing agent than bromine so chlorine oxidizes bromide ions.

Example

chlorine + bromide $\rightarrow$ chloride + bromine

$$Cl_2(aq) + 2Br^-(aq) \rightarrow 2Cl^-(aq) + Br_2(aq)$$

The **half-equations** for this reaction are

$$Cl_2(aq) + 2e^- \rightarrow 2Cl^-(aq) \text{ (reduction)}$$

Chlorine is reduced to chloride.

$$2Br^-(aq) \rightarrow + Br_2(aq) + 2e^- \text{(oxidation)}$$

Bromide is oxidized to bromine.

Bromine is a more powerful oxidizing agent than iodine so bromine oxidizes iodide ions.

Example

bromine + iodide $\rightarrow$ bromide + iodine

$$Br_2(aq) + 2I^-(aq) \rightarrow 2Br^-(aq) + I_2(aq)$$

The half-equations for this reaction are

$$Br_2(aq) + 2e^- \rightarrow 2Br^-(aq) \text{ (reduction)}$$

Bromine is reduced to bromide.

$$2I^-(aq) \rightarrow I_2(aq) + 2e^- \text{(oxidation)}$$

Iodide is oxidized to iodine.

Worked example

An aqueous solution of bromine is mixed with a solution of potassium iodide.

a Give the half-equation for the oxidation of iodide to iodine.

b Give the half-equation for the reduction of bromine to bromide

Answer

a $2I^-(aq) \rightarrow I_2(aq) + 2e^-$

b $Br_2(aq) + 2e^- \rightarrow 2Br^-(aq)$

Notice that iodine and bromine both exist as molecules.

Trends in the reducing ability of halide ions

- The halogens are good oxidizing agents
- The oxidizing ability of the halogens decreases down the group.

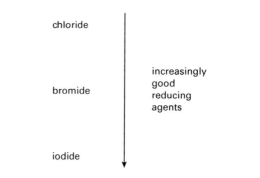

chlorine

bromine

oxidizing ability
of the halogens
decreases

iodine

- A reducing agent is a species which reduces another substance by adding electrons to it.
- Reducing agents are themselves oxidized during the reaction.
- Down the group the halide ions become increasingly good reducing agents.
  - As a result iodide ions are the strongest reducing agent. Iodide ions are easiest to oxidize.

chloride

bromide

increasingly
good
reducing
agents

iodide

Questions

1 Why does the boiling point of halogens increase down the group?
2 Explain what 'oxidizing ability' means
3 Why is chlorine a good oxidizing agent?

OBJECTIVES

By the end of this section you should be able to:

○ *recall the different products formed when sodium halide salts are reacted with sulfuric acid*

○ *understand how acidified $AgNO_3$ solution is used to identify fluoride, chloride, bromide, and iodide ions*

○ *recall the trend in solubility of silver halides in ammonia*

○ *recall how chlorine reacts with water and why chlorine is used in water treatments*

○ *recall how chlorine reacts with sodium hydroxide*

Sodium halide salts and sulfuric acid

• **Halide** ions are good **reducing agents**.

• Their strength as reducing agents increases down the group.

   ℗ As a result iodide ions are the strongest reducing agents. Iodide ions are easiest to oxidize.

This trend is illustrated by the reaction between sodium halide salts and concentrated sulfuric acid.

• Concentrated sulfuric acid is an oxidizing agent.

Sodium chloride

$$NaCl + H_2SO_4 \rightarrow HCl + NaHSO_4$$

The sulfuric acid reacts as an acid.

The HCl which is produced cannot reduce the sulfuric acid, so no redox reaction takes place.

Sodium bromide

$$NaBr + H_2SO_4 \rightarrow HBr + NaHSO_4$$

The sulfuric acid reacts as an acid.

The HBr which is produced is a strong enough reducing agent to reduce some of the sulfuric acid to sulfur dioxide.

$$2HBr + H_2SO_4 \rightarrow Br_2 + SO_2 + 2H_2O$$

The bromide ions are oxidized to bromine.

Sodium iodide

$$NaI + H_2SO_4 \rightarrow HI + NaHSO_4$$

The sulfuric acid reacts as an acid.

The HI which is produced is a strong reducing agent. It reduces the sulfuric acid to a mixture of reduced products.

$$2HI + H_2SO_4 \rightarrow I_2 + SO_2 + 2H_2O$$
$$6HI + H_2SO_4 \rightarrow 3I_2 + S + 4H_2O$$
$$8HI + H_2SO_4 \rightarrow 4I_2 + H_2S + 4H_2O$$

Worked example

Concentrated sulfuric acid reacts with sodium chloride to produce hydrogen chloride.

a Give the equation for this reaction.

b State the oxidation number of sulfur in concentrated sulfuric acid, H_2SO_4, sulfur dioxide, SO_2, and hydrogen sulfide, H_2S.

c Sulfur dioxide is produced when concentrated sulfuric acid is reacted with sodium bromide.

Hydrogen sulfide is produced when concentrated sulfuric acid is reacted with sodium iodide.

Explain how this shows that iodide ions are stronger reducing agents than bromide ions.

Answer

a $NaCl + H_2SO_4 \rightarrow HCl + NaHSO_4$

b In sulfuric acid it is 6+, in sulfur dioxide it is 4+, and in hydrogen sulfide it is 2−.

c Iodide ions are stronger reducing agents because they reduce sulfur from 6+ to 2− while bromide ions only reduce sulfur from 6+ to 4+.

Identifying halide ions

We can identify halide ions using **acidified silver nitrate solution**.

• Fluoride ions form silver fluoride which is a soluble salt so no precipitate is seen.

• The other halide ions form insoluble salts. The colour of the silver salt can be used to identify the halide.

halide ion	salt formed	colour of the precipitate
chloride	AgCl	white
bromide	AgBr	cream
iodide	AgI	yellow

• The solubility of the silver halides in ammonia can be used to confirm the identity of the halide ions.

silver halide	solubility in ammonia
AgCl	Dissolves in dilute ammonia solution.
AgBr	Does not dissolve in dilute ammonia solution but does dissolve in concentrated ammonia solution.
AgI	Does not dissolve even in concentrated ammonia solution.

Chlorine and water treatments

How does chlorine react with water?

Chlorine reacts with water to form two acids:

- chloric(I) acid
- hydrogen chloride.

$$Cl_2(aq) + H_2O(l) \rightleftharpoons HClO(aq) + HCl(aq)$$

The oxidation state of chlorine in

- Cl_2 is 0
- HClO is 1+
- HCl is 1−

Notice that this a disproportionation reaction.

The chlorine is simultaneously oxidized (from 0 to 1+) and reduced (from 0 to 1−).

High levels of micro-organisms in unchlorinated water can cause devastating outbreaks of diseases such as cholera. Sodium chlorate(I) is also used

Chloric(I) acid ionizes to form hydrogen ions and chlorate(I) ions.

$$HClO(aq) \rightleftharpoons H^+(aq) + ClO^-(aq)$$

Chloric(I) acid is is much better at killing micro-organisms than chlorate(I) ions so the pH of the water should be kept lower than 8 to ensure that there is a high concentration of chloric(I) acid which makes the chlorination of water more efficient and the water safer to drink.

Problems with using chlorine

There are some concerns about the use of chlorine in water treatments.

- Chlorine is a toxic gas.
- Chlorine also reacts with organic compounds to form substances which may be hazardous to humans although there is insufficient evidence to prove this link.

However, we continue to use chlorine because the health benefits of using chlorine far outweigh any possible problems it may cause.

The reaction of chlorine with sodium hydroxide

Chlorine reacts with cold, dilute sodium hydroxide to form sodium chloride, sodium chlorate(I), and water.

This is an example of **disproportionation**. The chlorine is simultaneously oxidized and reduced.

Sodium chlorate(I) is used to make household **bleaches**.

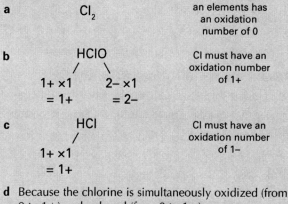

Worked example

The reaction between chlorine and water can be summed up by the equation.

$$Cl_2(aq) + H_2O(l) \rightleftharpoons HClO(aq) + HCl(aq)$$

Give the oxidation number of chlorine in:

a Cl_2 **b** HClO   **c** HCl

d Explain why the reaction between chlorine and water can be described as a disproportionation reaction.

Answer

a Cl_2 an elements has an oxidation number of 0

b   HClO

1+ ×1   2− ×1
= 1+   = 2−

Cl must have an oxidation number of 1+

c   HCl

1+ ×1
= 1+

Cl must have an oxidation number of 1−

d Because the chlorine is simultaneously oxidized (from 0 to 1+) and reduced (from 0 to 1−).

Worked example

State how you could prove that a solution contained chloride, Cl^-, ions.

Answer

Add acidified silver nitrate solution and you would see a white precipitate of silver chloride. This precipitate would be soluble in dilute ammonia solution.

HSW: Chlorine in the UK

In the UK very low levels of chlorine are added to drinking water to reduce levels of micro-organisms to acceptable levels.

Questions

1 What is the trend in the strength of the ability of halide ions to reduce substances?

2 Why are brown fumes seen when sodium bromide is added to concentrated sulfuric acid? Use equations to explain your answer.

3 How could you identify bromide ions?

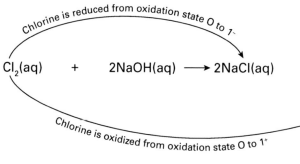

Chlorine is reduced from oxidation state 0 to 1⁻

$$Cl_2(aq) + 2NaOH(aq) \longrightarrow 2NaCl(aq) + NaOCl(aq) + H_2O(l)$$

Chlorine is oxidized from oxidation state 0 to 1⁺

2.10 Trends in group 2, the alkaline earth metals

Introducing the alkaline earth metals

- Group 2 is found in the s block of the periodic table.
- The group includes magnesium, calcium, strontium, and barium.

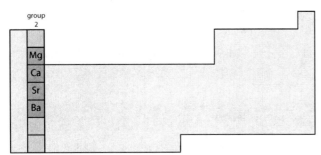

The position of group 2 in the periodic table.

- They are all reactive metals which are quite soft and have relatively low densities.
- Group 2 metals have two electrons in their outer shell. They react to form ions that have a 2+ charge.

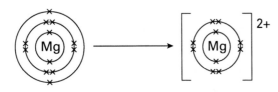

Magnesium atoms react to form magnesium ions, which have a 2+ charge.

Trends in atomic radius

element	atomic radius (nm)
Mg	0.145
Ca	0.194
Sr	0.219
Ba	0.253

- Down the group atoms have an extra shell of electrons and become larger.
  - As a result the **atomic radii** of the elements increase down the group.

Trends in ionization energy

- Down the group the outer electrons are lost more easily.
- Although the number of protons increases down the group, the atomic radii also increase so the distance between the nucleus and the outer electrons increases.
- There is also more shielding by the inner electrons.
  - As a result the value for the first ionization energy decreases down the group.

element	first ionization energy (kJ mol^{-1})
Mg	738
Ca	590
Sr	550
Ba	503

- The reactivity increases down the group.
- Group 2 elements are strong **reducing agents** because they can lose electrons quite easily.
- Down the group the elements become stronger reducing agents.

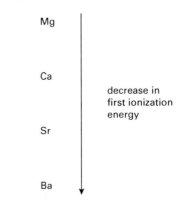

Trends in melting points

Group 2 elements have reasonably high melting points due to **metallic bonding**.

element	melting point (°C)
Mg	649
Ca	839
Sr	769
Ba	729

Metallic bonding

- In metals the electrons in the outer shell of atoms are **delocalized**. This leads to positive metal ions and negatively charged delocalized electrons.

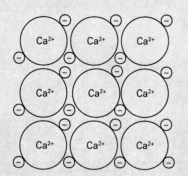

The metallic bonding in calcium metal.

- Metallic bonding is the electrostatic attraction between the positive ions and the negative delocalized electrons.
- Down the group the size of the metal ions increases so the strength of the metallic bonding decreases.
- Less energy is required to overcome the forces of attraction.

  As a result melting points generally decrease down the group.

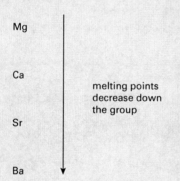

melting points decrease down the group

How do Group 2 elements react with water?

Group 2 metal atoms are good reducing agents. Their reaction with water is an example of a **redox** reaction.

- The group 2 metal atom is oxidized.

$$M \rightarrow M^{2+} + 2e^-$$

- The hydrogen in water is reduced.

Magnesium reacts slowly with water to form magnesium hydroxide and hydrogen.

$$Mg(s) + 2H_2O(l) \rightarrow Mg(OH)_2(aq) + H_2(g)$$

Magnesium hydroxide is sparingly soluble.
Magnesium reacts quickly with steam.

$$Mg(s) + H_2O(g) \rightarrow MgO(s) + H_2(g)$$

Calcium, strontium and barium all react vigorously with water to form a metal hydroxide and hydrogen.

Example

$$Ca(s) + 2H_2O(l) \rightarrow Ca(OH)_2(aq) + H_2(g)$$

A solution of calcium hydroxide looks cloudy as it is only slightly soluble in water and the undissolved calcium hydroxide forms a white suspension.

Worked example

Calcium and strontium react with cold water.

a What would you see as calcium reacts with water?

b Write a balanced symbol equation for the reaction between strontium and water.

Answer

a Fizzing or bubbles (as hydrogen is formed); a white suspension (of calcium hydroxide forming); metal gets smaller (as it reacts).

b $Sr(s) + 2H_2O(l) \rightarrow Sr(OH)_2(aq) + H_2(g)$

Trends in the way that group 2 elements react with water

Down the group the elements react more vigorously with water.

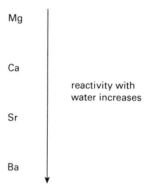

reactivity with water increases

Worked example

a Which of these metal hydroxides is most soluble in water?

Magnesium hydroxide, calcium hydroxide, strontium hydroxide, barium hydroxide.

b Which of these metals reacts most vigorously with water?

Magnesium, calcium, strontium, barium.

Answer

a barium hydroxide (solubility of group 2 hydroxides increases down the group)

b barium (elements react more vigorously with water down the group)

Questions

1 Describe the trend in atomic radii down group 2 of the periodic table.

2 Describe the trend in ionization energy down group 2 of the periodic table.

3 Give the symbol equation for the reaction between strontium and water.

2.11 Trends in the chemical properties of group 2 metals

OBJECTIVES

By the end of this section you should be able to:

○ recall the relative solubilities of group 2 hydroxides

○ recall how magnesium hydroxide is used in medicine and calcium hydroxide is used in agriculture

○ recall the relative solubilities of group 2 sulfates

○ understand how to identify sulfate ions using acidified barium chloride solution

○ recall how barium sulfate is used in medicine

Relative solubilities of group 2 hydroxides

group 2 hydroxide	solubility (g per 100 cm³ of water)	solubility
$Mg(OH)_2$	0.0012	
$Ca(OH)_2$	0.113	⬇
$Sr(OH)_2$	0.410	
$Ba(OH)_2$	2.57	

- The **solubility** of group 2 hydroxides increases down the group.
- Magnesium hydroxide, $Mg(OH)_2$ is only sparingly soluble in water.
  - Ⓔ As a result when magnesium hydroxide is added to water it forms a white suspension which is weakly **alkaline**.
- Calcium hydroxide, $Ca(OH)_2$, and strontium hydroxide, $Sr(OH)_2$, are slightly soluble.
  - Ⓔ As a result when they added to water they form alkaline suspensions.
- Barium hydroxide, $Ba(OH)_2$ is quite soluble.
  - Ⓔ As a result when barium hydroxide is added to water it forms an alkaline solution. The solution is more alkaline because there is a higher concentration of hydroxide, OH^-, ions from the barium hydroxide.

Worked example

a Describe what you would see when magnesium hydroxide is added to water.

b Why is a solution of strontium hydroxide more strongly alkaline than a solution of magnesium hydroxide?

Answer

a A white suspension because magnesium hydroxide is only slightly soluble in water.

b Strontium hydroxide is more strongly alkaline because there is a higher concentration of hydroxide ions from the strontium hydroxide because it is more soluble than magnesium hydroxide.

Addition of sodium hydroxide solution

If we add **sodium hydroxide** solution to a solution of a group 2 salt we would form a **precipitate** of the group 2 metal hydroxide.

Example

$$Ca^{2+}(aq) + 2OH^-(aq) \rightarrow Ca(OH)_2(s)$$

As barium hydroxide is more soluble it only forms a faint white precipitate.

Worked example

Which of these metal hydroxides is most soluble in water?

$Mg(OH)_2$, $Ca(OH)_2$, $Sr(OH)_2$, $Ba(OH)_2$.

Answer

$Ba(OH)_2$, solubility increases down the group.

Uses of magnesium hydroxide in medicine

A solution of magnesium hydroxide called milk of magnesia is used as an antacid medicine.

It is used to treat indigestion and heartburn.

Indigestion can be caused when the stomach produces too much hydrochloric acid.

Heartburn is caused when stomach acid is pushed into the oesophagus.

Milk of magnesia relieves these conditions by neutralizing this excess acid.

$$Mg(OH)_2(s) + 2HCl(aq) \rightarrow MgCl_2(aq) + 2H_2O(l)$$

Worked example

The stomach produces hydrochloric acid.

A solution of magnesium hydroxide can be used to treat indigestion and heartburn.

a How is heartburn caused?

b Give the balanced symbol equation for the reaction between magnesium hydroxide and hydrochloric acid.

Answer

a It is caused when stomach acid is pushed into the oesophagus.

b $Mg(OH)_2(s) + 2HCl(aq) \rightarrow MgCl_2(aq) + 2H_2O(l)$

Use of calcium hydroxide in agriculture

Calcium carbonate is found in many rocks such as limestone. When heated strongly, the calcium carbonate reacts to form calcium oxide and carbon dioxide. This is a thermal decomposition reaction.

$$CaCO_3(s) \rightarrow CaO(s) + CO_2(g)$$

The calcium oxide that is made can be reacted with water to form calcium hydroxide, $Ca(OH)_2$.

$$CaO(s) + H_2O(l) \rightarrow Ca(OH)_2(s)$$

Crops normally only grow well in soil which has a certain pH.

Acids from acid rain and fertilizers lower the pH of soil so farmers **'lime'** soils using powdered limestone, calcium oxide, or calcium hydroxide to increase the pH of the soil so that their crops grow better.

Worked example

a Why is 'lime' applied to fields?

b Use balanced symbol equations to explain how calcium hydroxide is made.

Answer

a Soils often become acidic because of acid rain and fertilizers but most crops only grow well in soils which have a certain pH.

b First calcium carbonate is heated to form calcium oxide.

$$CaCO_3(s) \rightarrow CaO(s) + CO_2(g)$$

Then calcium oxide is reacted with water to form calcium hydroxide.

$$CaO(s) + H_2O(l) \rightarrow Ca(OH)_2(s)$$

Relative solubilities of group 2 sulfates

group 2 sulfate	solubility (g per 100 cm³ of water)	solubility
$MgSO_4$	25.5	
$CaSO_4$	0.24	
$SrSO_4$	0.013	
$BaSO_4$	0.00022	

- The solubility of group 2 sulfates decreases down the group.
- Magnesium sulfate is very soluble while barium sulfate is very insoluble.

Worked example

Which of these metal sulfates is most soluble in water?

Magnesium sulfate, calcium sulfate, strontium sulfate, barium sulfate.

Answer

Magnesium sulfate is most soluble (solubility decreases down the group).

How to identify sulfate ions

We can identify sulfate ions by adding a solution of **acidified barium chloride** (which contains barium ions) to a solution that contains sulfate ions. A dense white precipitate of barium sulfate is formed.

$$Ba^{2+}(aq) + SO_4^{2-}(aq) \rightarrow BaSO_4(s)$$

Worked example

a Describe how you could use barium chloride solution to prove that a solution contained sulfate ions.

b Give the equation for the reaction between barium ions and sulfate ions. Include state symbols.

Answer

a When you add barium chloride solution to the solution containing sulfate ions you would see a white precipitate of barium sulfate.

b $Ba^{2+}(aq) + SO_4^{2+}(aq) \rightarrow BaSO_4(s)$

Use of barium sulfate in medicine

- Barium compounds are toxic but barium sulfate is regularly used in medicine.
- Barium sulfate is very insoluble in water.
- In addition X-rays cannot pass through barium sulfate.

Patients suffering digestive problems may be given a 'barium meal' to drink or a barium enema. These both contain a barium sulfate suspension.

The patient is then X-rayed and the doctors can see any abnormalities in the digestive system without having to operate on the patient.

The insoluble barium sulfate eventually passes out of the body.

Worked example

Explain why when calcium is placed in sulfuric acid it reacts for a few seconds and then the reaction stops.

Answer

Sulfuric acid contains sulfate ions. When calcium reacts with sulfuric acid the salt calcium sulfate forms. Calcium sulfate is insoluble. This forms a protective layer around the calcium and prevents any further reaction from occurring.

Questions

1 Why is a solution of strontium hydroxide more alkaline than a solution of magnesium hydroxide?

2 How is calcium hydroxide used in agriculture?

3 How can we prove that sulfate ions are present in a solution?

2.12 Extraction of metals

By the end of this section you should be able to:

○ recall that metals are found in ores which are usually metal oxides or sulfides

○ recall how metal sulfides are converted to metal oxides

○ explain why the extraction of metals from metal oxides is an example of reduction

○ understand why carbon and carbon monoxide are used in the extraction of iron, manganese, and copper

○ understand how aluminium is extracted from bauxite, how titanium is extracted from TiO_4, and how tungsten is extracted form WO_3

○ recall the advantages and disadvantages of recycling scrap metals

○ recall the environmental advantages of using scrap iron to extract copper from aqueous solutions formed from low grade ores

Metal ores

Most rocks contain a mixture of **minerals**. Minerals are chemical compounds that have a particular crystal structure. A rock that contains a mineral in high enough concentrations that it is economically worthwhile to extract the mineral from the rock is called an **ore**.

mineral	formula of the mineral	metal extracted from the ore
haematite	Fe_2O_3	iron
pyrolusite	MnO_2	manganese
chalcopyrite	$CuFeS_2$	copper
rutile	TiO_2	titanium
bauxite	Al_2O_3	aluminium
scheelite	$CaWO_4$	tungsten

Most metal ores are oxides or sulfides.

Converting metal sulfides to metal oxides

It is hard to extract a metal from a metal sulfide so an ore that contains a metal sulfide is normally first converted into a metal oxide.

This is done by crushing the ore to increase its surface area and then roasting it in air.

Example

The main ore of copper, chalcopyrite, $CuFeS_2$, is roasted in air to form copper oxide, CuO.

$$2CuFeS_2(s) + 6\tfrac{1}{2} O_2(g) \rightarrow 2CuO(s) + Fe_2O_3(s) + 4SO_2(g)$$

If the SO_2 was released into the atmosphere it could cause acid rain, so it is important that it does not get into the environment.

In fact, the sulfur dioxide can be used to make sulfuric acid. This has both environmental and economic benefits.

Reduction reactions

Metals are extracted from metal oxides by **reduction** reactions.

The method of reduction depends on the reactivity of the metal.

The main methods of extracting metals are:

- reducing the ore using carbon or carbon monoxide
- reducing the molten ore by **electrolysis**
- reducing the metal oxide using a more reactive metal

Carbon and carbon monoxide

- Carbon in the form of coke is used to reduce metal oxides whenever possible.
- **Coke** is a very cheap reducing agent.
- In the conditions used some coke may be converted to carbon monoxide
- Carbon monoxide is also a reducing agent.
- Coke and carbon monoxide are used to extract fairly unreactive metals such as iron, manganese, and copper.

Extraction of iron

Iron is extracted from its ore **haematite**, which contains iron(III) oxide, Fe_2O_3.

- Inside the blast furnace the coke burns to form carbon dioxide.

$$C(s) + O_2(g) \rightarrow CO_2(g)$$

- In the conditions used some carbon dioxide reacts with more carbon to form carbon monoxide.

$$CO_2(g) + C(s) \rightarrow 2CO(g)$$

- The carbon monoxide reduces the iron(III) oxide to iron and carbon dioxide.

$$Fe_2O_3(s) + 3CO(g) \rightarrow 2Fe(l) + 3CO_2(g)$$

- If the liquid iron is allowed to cool it solidifies to form **cast iron**.

Extraction of manganese

The main ore of manganese is pyrolusite, which contains manganese(IV) oxide, MnO_2.

- The manganese(IV) oxide is first reduced to manganese(II) oxide by carbon monoxide.

$$MnO_2(s) + CO(g) \rightarrow MnO(s) + CO_2(g)$$

- The manganese(II) oxide is then further reduced to manganese.

$$MnO(s) + C(s) \rightarrow Mn(l) + CO(g)$$

Extraction of copper

Copper can be extracted from its ore chalcopyrite, which contains copper(II) oxide, CuO.

- Carbon reduces copper(II) oxide to copper.

$$CuO(s) + C(s) \rightarrow Cu(l) + CO(g).$$

Extraction of aluminium

Aluminium is a fairly reactive metal which is extracted form its ore bauxite by **electrolysis**. Bauxite contains aluminium oxide, Al_2O_3.

- Bauxite has a very high melting point and the energy costs of heating bauxite to this temperature would be very high.

Instead, a second ore of aluminium, cryolite, Na_3AlF_6, which has a lower melting point is used.

- The cryolite is heated until it melts and then bauxite is dissolved in the molten cryolite.
- Eectrolysis is then used to reduce the aluminium ions.
  At the **anode** (positive electrode)

$$2O^{2-} \rightarrow O_2 + 4e^-$$

At the **cathode** (negative electrode)

$$Al^{3+} + 3e^- \rightarrow Al$$

- The anode is made of graphite which is a form of carbon.
- The oxygen produced at the anode reacts with the electrode to form carbon dioxide. So the electrodes must be replaced periodically.

$$O_2(g) + C(s) \rightarrow CO_2(g)$$

Extraction of titanium

Titanium is a fairly reactive metal which is extracted from its ore rutile by reduction using a more reactive metal. Rutile contains titanium(IV) oxide, TiO_2.

- First the titanium(IV) oxide is converted to titanium(IV) chloride.

$$TiO_2(s) + 2C(s) + 2Cl_2(g) \rightarrow TiCl_4(g) + 2CO(g)$$

- Then the titanium(IV) chloride is reduced by a more reactive metal such as sodium or magnesium.

$$TiCl_4(g) + 2Mg(l) \rightarrow Ti(s) + 2MgCl_2(l)$$

- As the sodium or magnesium used to reduce the titanium must be extracted by electrolysis this makes titanium a very expensive metal to produce.

Extraction of tungsten

Tungsten is a fairly unreactive metal.

- First the tungsten ore is converted into tungsten(VI) oxide.
- Then tungsten is extracted from tungsten(VI) oxide by reduction with hydrogen.

$$WO_3(s) + 3H_2(g) \rightarrow W(s) + 3H_2O(g)$$

- The use of hydrogen as a reducing agent has some risks.
- Hydrogen is a very **flammable** gas which forms an explosive mixture with air.
- Also because hydrogen is colourless and odourless hydrogen leaks are very difficult to detect.

HSW: Copper and acid rain

Extracting copper from its main ore produces SO_2.

HSW: Recycling scrap metals

We have limited supplies of metal ores so it is important that we recycle metals so that we do not run out of this valuable resource.

- *Recycling metals reduces the amount of metal ore that must be mined.*
- *This protects the environment where the mine would have been built and also the area around the mine.*
- *The area around a mine can be damaged when acidic solutions dissolve toxic chemicals from the mine and then contaminate nearby rivers and streams.*
- *Extracting metals from their ores requires a lot of energy. Recycling metals uses much less energy.*
  - *As a result fewer fossil fuels need to be burnt.*
- *This has economic advantages and by reducing the amount of carbon dioxide that is produced and released into the atmosphere it also has environmental advantages.*

However, recycling metals can have disadvantages for some people.

- *Many mines are found in poor parts of the world, where there are few employment opportunities.*
- *If these mines are closed local people will lose their jobs.*

Using scrap iron to extract copper

Copper has been used since ancient times. Today, we have used up most **high-grade** ores so we must extract copper from lower grade ores.

- Enormous amounts of rock must be mined to extract enough copper and this has a massive environmental impact.
- Scientists have developed a new method of extracting copper from these low-grade ores. First the copper is leached from the ore to form a solution that contains Cu^{2+} ions.
- Scrap iron is then used to recover the copper from these solutions.

$$Cu^{2+}(aq) + Fe(s) \rightarrow Cu(s) + Fe^{2+}(aq)$$

Reactivity of metals

Sodium, Na
Magnesium, Mg
Aluminium, Al
Titanium, Ti
Manganese, Mn
Iron, Fe
Tungsten, W

reactivity increases

Questions

1 What is an 'ore'?
2 How is tungsten extracted from WO_3?
3 Give the equation for the reaction in which scrap iron is used to recover copper from copper solutions.

By the end of this section you should be able to:

○ reproduce the reaction mechanisms for the reaction between methane and chlorine

○ recall how chloroalkanes and chlorofluoroalkanes are used

○ explain the benefits of the ozone formed in the upper atmosphere

○ recall how chlorine atoms catalyse the decomposition of ozone and the impact of this on the ozone layer

○ recall how chlorine atoms form in the upper atmosphere

○ describe the legislation to ban CFCs and how scientists have developed alternatives

Free radicals

- **Free radicals** are very reactive species. They are species which have an unpaired electron.
- They are formed when a covalent bond breaks so that one electron is transferred to each of the atoms.
- This is called **homolytic fission** and forms two free radicals.

$$A-B \rightarrow A\bullet + B\bullet$$

Notice that the dot, which represents the unpaired electron, is written next to the atom which has the unpaired electron.

Worked example

a Explain what a free radical is.

b Write an equation to show how a chlorine molecule can form chlorine free radicals.

Answer

a A free radical is a species which has an unpaired electron.

b $Cl_2 \rightarrow 2Cl\bullet$

The free radical substitution of methane

Methane reacts with chlorine by a free radical substitution reaction.

We can sum up this reaction using the equation

$$CH_4 + Cl_2 \rightarrow CH_3Cl + HCl$$

Initiation

- UV light provides the energy required for the homolytic fission of the chlorine molecule.

$$Cl_2 \rightarrow 2Cl\bullet$$

Propagation

- Next the chlorine free radical reacts with methane to form hydrogen chloride and a methyl free radical.

$$Cl\bullet + CH_4 \rightarrow HCl + \bullet CH_3$$

- Then the methyl free radical reacts to form chloromethane and a chlorine free radical.

$$\bullet CH_3 + Cl_2 \rightarrow CH_3Cl + Cl\bullet$$

Notice that as the free radical reacts it forms a new molecule and another free radical.

Other products include

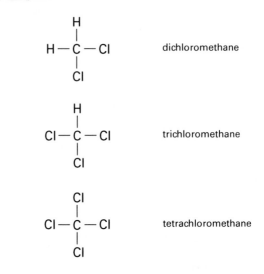

dichloromethane

trichloromethane

tetrachloromethane

Termination

In the termination step two free radicals react together to form a new molecule.

$$Cl\bullet + Cl\bullet \rightarrow Cl_2$$
$$\bullet CH_3 + Cl\bullet \rightarrow CH_3Cl$$
$$\bullet CH_3 + \bullet CH_3 \rightarrow C_2H_6$$

Notice that the three steps in the free radical substitution of methane are initiation, propagation, and termination.

- In the **initiation** step free radicals are made.
- In the **propagation** steps free radicals react with molecules to form new molecules and free radicals.
- In the **termination** steps two free radicals join together. A variety of new molecules are formed.

Worked example

In the presence of UV light methane reacts with chlorine.

a Why does the reaction require the presence of UV light?

b State the formula of a product of the reaction which does not contain chlorine.

Explain how this product is formed.

Answer

a The UV light is required to split the chlorine molecule into two chlorine free radicals.

b C_2H_6. It is formed by the termination step involving two methyl free radicals.

$$\bullet CH_3 + \bullet CH_3 \rightarrow C_2H_6$$

Uses of chloroalkanes and chlorofluoroalkanes

- **Chloroalkanes** are alkane molecules which have one or more hydrogen atoms substituted by chlorine atoms.
- Chloroalkanes are good solvents.
- **Chlorofluoroalkanes**, CFCs, are alkanes that have hydrogen atoms substituted by chlorine and fluorine atoms.
- CFCs have been used as propellants in aerosols and as refrigerants.
- CFCs can escape into the Earth's atmosphere where they can damage the ozone layer.

Benefits of the ozone layer

- **Ozone**, O_3, is formed in the upper atmosphere.
- UV light from the Sun provides the energy required to break the covalent bond in an oxygen molecule.

$$O_2 \rightarrow \bullet O \bullet + \bullet O \bullet$$

- The oxygen free radicals then react with oxygen molecules to form ozone.

$$\bullet O \bullet + O_2 \rightarrow O_3$$

- Ozone helps to filter out harmful UV radiation.

This is important because over-exposure to UV radiation can cause skin cancers, cataracts, and damage to crops.

Worked example

a Why is ozone useful to life on Earth?

b Explain using balanced equations how ozone can form.

Answer

a Ozone helps to filter out harmful UV light.

b UV light can break the bond in oxygen molecules to form oxygen free radicals.

$$O_2 \rightarrow \bullet O \bullet + \bullet O \bullet$$

These free radicals can then react with oxygen molecules to form ozone.

$$\bullet O \bullet + O_2 \rightarrow O_3$$

Decomposition of ozone

- In the upper atmosphere the CFC molecules decompose to form free radicals.

$$CCl_3F \rightarrow \bullet CCl_2F + Cl \bullet$$

- The chlorine free radical then removes ozone forming the chlorine oxide free radical and oxygen.

$$Cl \bullet + O_3 \rightarrow ClO \bullet + O_2$$

Notice how the chlorine oxide free radical can also react with ozone.

$$ClO \bullet + O_3 \rightarrow 2O_2 + Cl \bullet$$

Notice that a chlorine free radical is produced in this reaction. This can decompose more ozone.

- Many old refrigerators and freezers contain CFCs. When they are disposed of the CFCs must first be removed to prevent the gases escaping into the atmosphere.

Legislation and new compounds

- Scientists have helped us to understand how the hole in the ozone layer has formed and the legislation to ban CFCs has been widely supported by scientists.
- Scientists have also developed alternative compounds including hydrochlorofluorocarbons, HCFCs, which contain less chlorine and HFCs which do not contain chlorine and are not thought to affect the atmosphere.

Worked example

a In the past CFCs were widely used. State a use of CFCs.

b Explain why the use of CFCs has been greatly reduced.

c Today HFCs have widely replaced CFCs in many applications. Why are HFCs not thought to affect the Earth's atmosphere?

Answer

a They were widely used as propellants in aerosols and as refrigerants.

b CFCs decompose to form free radicals.

$$CCl_3F \rightarrow \bullet CCl_2F + Cl \bullet$$

These free radicals then react with ozone.

$$Cl \bullet + O_3 \rightarrow ClO \bullet + O_2$$

$$ClO \bullet + O_3 \rightarrow 2O_2 + Cl \bullet$$

c HFCs do not contain chlorine.

HSW: Ozone and CFCs

Scientists first discovered that the level of ozone in the upper atmosphere was decreasing. They discovered that the 'hole' in the ozone layer is caused by CFCs which had escaped into the atmosphere.

Questions

1 How are free radicals formed?
2 Name the type of reaction by which methane reacts with chlorine.
3 How are CFCs used?

2.14 Haloalkanes 2

OBJECTIVES

By the end of this section you should be able to:

○ recall that haloalkanes contain polar bonds

○ describe how haloalkanes can be attacked by OH⁻, CN⁻, and NH₃

○ understand the nucleophilic substitution of primary haloalkanes

○ describe how the strength of the carbon–halogen bond influences the rate of hydrolysis of haloalkanes

○ recall that haloalkanes can react by concurrent substitution and elimination

○ describe how these reactions can be used in organic synthesis

More about haloalkanes

The haloalkanes, $C_nH_{2n+1}X$, are also known as **halogenoalkanes** and **alkyl halides**.

They are alkanes in which one hydrogen atom has been substituted by a halogen atom, X, where X can be fluorine, chlorine, bromine, or iodine.

Haloalkanes can be classified as **primary**, **secondary**, or **tertiary**.

Primary haloalkanes

• The halogen is bonded to a carbon which is bonded to just one other carbon atom.

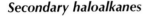
1-bromopropane

Secondary haloalkanes

• The halogen is bonded to a carbon which is bonded to two other carbon atoms.

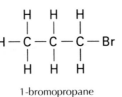
2-bromopropane

Tertiary haloalkanes

• The halogen is bonded to a carbon which is bonded to three other carbon atoms.

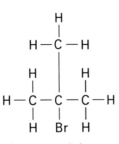
2-bromo-2-methylpropane

Polar bonds

Most halogen atoms are more electronegative than carbon. This means that haloalkanes have polar C–X bonds. (Notice that X is used to represent a halogen atom.)

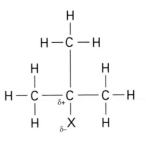

Nucleophiles

Nucleophiles such as OH⁻, CN⁻, and NH₃ have a lone pair of electrons which they can donate to another molecule to form a new covalent bond.

The nucleophilic substitution of primary haloalkanes

Hydroxide ions

Haloalkanes undergo **nucleophilic substitution** reactions with hydroxide ions to form alcohols.

This reaction can also be described as **hydrolysis**.

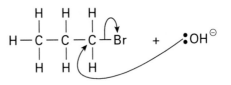

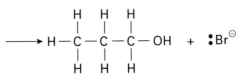

Cyanide ions

Haloalkanes undergo nucleophilic substitution reactions with **cyanide** ions to form nitriles.

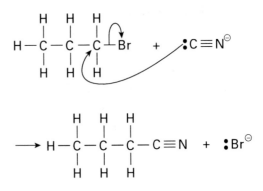

Ammonia molecules

Haloalkanes undergo nucleophilic substitution reactions with ammonia to form amines.

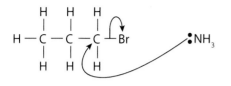

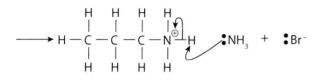

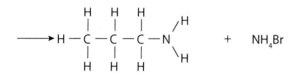

Strength of the C–X bond

The reactivity of haloalkanes depends on the halogen involved.

- **Polar bonds** are more **susceptible** to attack from nucleophiles than non-polar bonds.
- Iodoalkanes react faster with hydroxide ions than bromoalkanes or chloroalkanes.

The C–Cl bond has a very high bond enthalpy. This means that the bond is very strong and hard to break. The **activation energy** required to break the bond and start the reaction is so high that the reaction is very slow.

bond	bond enthalpy (kJ mol⁻¹)
C–Cl	338
C–Br	276
C–I	238

- Remember that the weaker the bond the faster the hydrolysis reaction.
  - As a result **bond enthalpy** is a more important factor than polarity.

Questions

1 What is a nucleophile?

2 Why do iodoalkanes react faster than chloroalkanes?

3 What is the organic product of the reaction between 2-bromopropane and hydroxide ions in anhydrous conditions?

The importance of conditions used

The reaction between haloalkanes and hydroxide ions can produce different products depending on the conditions chosen.

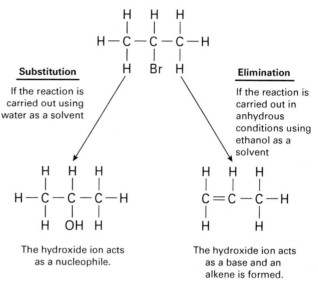

Substitution
If the reaction is carried out using water as a solvent

Elimination
If the reaction is carried out in anhydrous conditions using ethanol as a solvent

The hydroxide ion acts as a nucleophile.

The hydroxide ion acts as a base and an alkene is formed.

Mechanisms for the substitution reaction

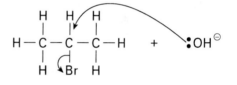

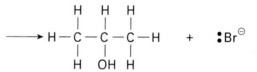

Mechanisms for the elimination reaction

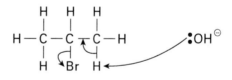

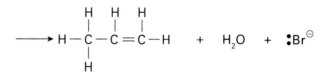

Using nucleophilic substitution reactions

- The nucleophilic substitution reactions of haloalkanes are used to introduce new functional groups to organic molecules.
- Using cyanide ions allows us to increase the carbon chain length.

OBJECTIVES

By the end of this section you should be able to:

○ recall that alkenes are unsaturated hydrocarbons that have a double covalent bond

○ recall that the carbon double bond is planar and that alkenes can show E-Z stereoisomerism owing to restricted rotation about the C=C double bond

○ draw E and Z isomers

○ understand the mechanisms by which alkenes react with Br_2, HBr, and H_2SO_4

○ recall that bromine can be used to test for unsaturation

Introducing alkenes

Alkenes are **unsaturated** hydrocarbons with the general formula C_nH_{2n}.

Alkenes have a C=C double bond which consists of a **sigma bond** and a **pi bond**. The pi bond forms above and below the axis of the carbon atoms.

Ethene is a **planar** (flat) molecule because of the C=C arrangement in alkene molecules.

The pi bond has a high electron density. Electrophiles can attack this high electron density.

As a result alkenes are more reactive than alkanes.

E-Z stereoisomers

Rotation of the C=C bond would require the pi bond to be broken.

This requires a lot of energy so there is restricted rotation about the C=C double bond.

As a result but-2-ene exists as two **stereoisomers**. As these isomers are not mirror images of each other they are called **diastereoisomers**.

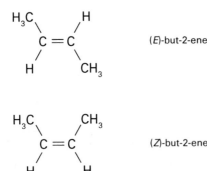

Notice that in the E form the methyl groups are arranged on opposite sides of the C=C bond. The E comes from the German word *entgegen* which means opposite.

In the Z form the methyl groups are arranged on the same side of the C=C bond. The Z comes from the German word *zusammen* which means together.

But-2-ene has diasteroeisomers because it has two different groups attached to each of the double bonded carbon atoms.

Reactions of alkenes

In addition reactions two species join together.

An **electrophile** is a species that can accept a pair of electrons to form a new covalent bond.

Electrophiles can attack the high electron density of the pi bond in alkenes.

As a result alkenes undergo electrophilic addition reactions.

Worked example

Define the term electrophile.

Answer

An electrophile is a species which can accept a pair of electrons to form a new covalent bond.

Reaction with bromine

Alkenes react with bromine to form dibromoalkanes.

Example

$$C_2H_4 + Br_2 \rightarrow C_2H_4Br_2$$

This is an electrophilic addition reaction.

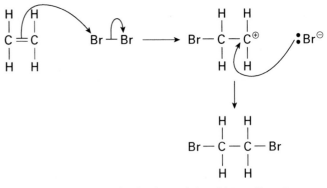

The reaction mechanism for the electrophilic addition of bromine to ethene.

Worked example

a Give the equation for the reaction between propene, C_3H_6, and bromine, Br_2, using structural formulae.

b What type of reaction is this?

Answer

a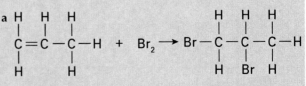

b electrophilic addition

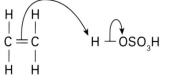

Reaction with hydrogen bromide

Alkenes also react with hydrogen bromide to form bromoalkanes.

Example

$$C_2H_4 + HBr \rightarrow C_2H_5Br$$

This is an electrophilic addition reaction.

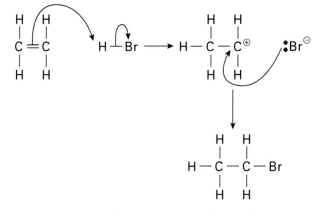

The reaction mechanism for the electrophilic addition of hydrogen bromide to ethene.

Worked example

a Give the equation for the reaction between ethene, C_2H_4, and HCl using structural formulae.

b What type of reaction is this?

Answer

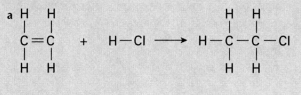

b electrophilic addition

Reaction with sulfuric acid

Alkenes react with sulfuric acid and then water to form alcohols.

Example

$$C_2H_4 + HOSO_3H \rightarrow CH_3CH_2(OSO_3H)$$

This is an electrophilic addition reaction.

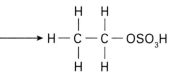

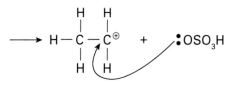

The reaction mechanism for the electrophilic addition of sulfuric acid to ethene.

Next the ethyl hydrogen sulfate is warmed with water to form ethanol.

$$CH_3CH_2(OSO_3H) + H_2O \rightarrow CH_3CH_2OH + H_2SO_4$$

Notice that the sulfuric acid used in the first part of the reaction is re-formed in the second part of the reaction.

Summary of the reactions of ethene

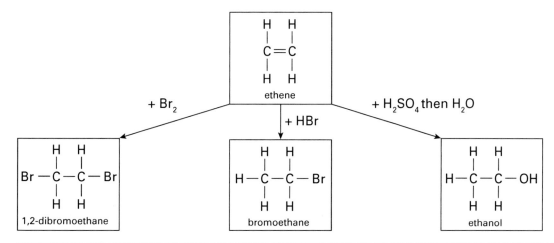

Questions

1 What are diastereoisomers?

2 What is an electrophile?

3 Name the type of reaction that takes place between bromine and ethene.

2.16 Alkenes 2

OBJECTIVES

By the end of this section you should be able to:

○ predict and explain the products of addition reactions to unsymmetrical alkenes

○ recall how alcohols are produced from alkenes industrially and the conditions used to convert ethene to ethanol

○ recall how addition polymers are made from alkenes

○ explain why poly(alkenes) are unreactive

○ recognize the repeating unit in a polyalkene

○ recall how poly(ethene) and poly(propene) are used and that poly(propene) is recycled

Unsymmetrical alkenes

Ethene is a **symmetrical** alkene. The groups attached to the C=C bond are the same.

$$
\begin{array}{cc}
H & H \\
| & | \\
C & = C \\
| & | \\
H & H
\end{array}
$$

When symmetrical alkenes such as ethene take part in addition reactions with **electrophiles** such as hydrogen bromide it does not matter how the hydrogen bromide adds across the carbon double bond because the product is always bromoethane.

However, this is not true for all alkene molecules.

Propene is an **unsymmetrical** alkene.

$$
\begin{array}{ccc}
H & H & H \\
| & | & | \\
C & = C & - C - H \\
| & & | \\
H & & H
\end{array}
$$

Notice that the groups attached to the carbon atoms in the C=C bond are different. This is significant when propene takes part in an addition reaction with an electrophile such as HBr. The HBr can add across the bond in two different ways, to produce 2-bromopropane or 1-bromopropane.

A helpful guide!

A useful guide for predicting the major product of an addition reaction with an unsymmetrical alkene is to use **Markovnikov's rule**.

- Markovnikov's rule states that the hydrogen goes to the carbon atom which has the most hydrogens already.

Why is 2-bromopropane the major product?

The major product for this reaction is 2-bromopropane.

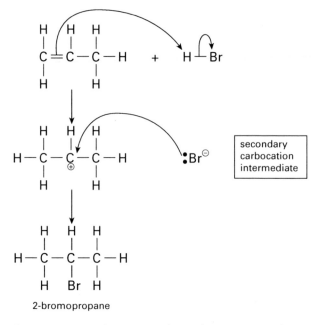

2-bromopropane

This reaction proceeds via a secondary carbocation intermediate.

This reaction proceeds via a secondary **carbocation** intermediate. This is more stable than a primary carbocation intermediate.

The minor product for this reaction is 1-bromopropane.

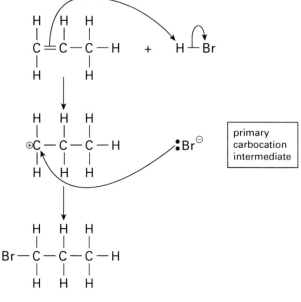

1-bromopropane

This reaction proceeds via a primary carbocation intermediate.

Producing alcohols

- Traditionally, alcohols have been produced by **fermentation**.

- Today, we can also produce alcohols by the **hydration** of alkenes.

- Alkenes are produced from crude oil.

  ⊛ As a result this is a non-renewable way of producing alcohols.

Ethanol is made from ethene by an addition reaction.

$$C_2H_4 + H_2O \rightarrow C_2H_5OH$$

Polymers

Lots of alkene molecules can be joined together to form **addition polymers**.

- The alkene molecules are called **monomers**.
- This is an addition **polymerization** reaction.

Many ethene molecules can join together to form the addition polymer poly(ethene).

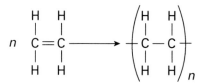

The formation of poly(ethene).

Notice that poly(alkenes) are **saturated** – they do not have C=C double bonds.

Many propene molecules can join together to form poly(propene).

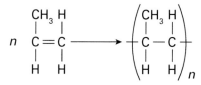

The formation of poly(propene).

The section drawn in brackets is called a **repeating unit**. The repeating unit is repeated thousands of times in each polymer molecule. The lines representing the covalent bonds between repeating units cross through the brackets around the repeating units.

Worked example

a Draw the structural formula of chloroethene.

b Chloroethene can be polymerized in a similar type of reaction to the polymerization of ethene to poly(ethene). Write the equation for the production of poly(chloroethene) from chloroethene. Clearly show the repeating unit of the polymer.

Answer

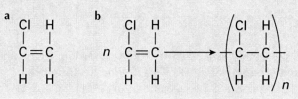

Why are poly(alkenes) unreactive?

- Alkenes have a sigma and a pi bond.
- The high electron density and easy accessibility of the pi bond means that alkenes can be attacked by electrophiles.
- Poly(alkenes) are saturated hydrocarbons.
  - As a result they are much less reactive than alkenes.

Using polymers

Poly(ethene)

- Poly(ethene) is a soft, flexible polymer.
- It is widely used to make plastic bags and bottles.
- Poly(ethene) does not have a sharp melting point.
- It melts over a range of temperatures because it contains a mixture of polymer chains that have different lengths.

Poly(propene)

- Poly(propene) is a tough, strong polymer.
- It is used to make buckets and crates.
- Plastics like poly(propene) can be recycled.
  - First the plastics are separated out.
  - Then the poly(propene) is shredded and then turned into granules.
  - These granules are then melted down and made into useful new items like flower pots.

Worked example

a Poly(tetrafluoroethene) is an expensive polymer. It is formed by an addition polymerization reaction between many tetrafluoroethene molecules. Draw the structural formula of tetrafluoroethene.

b Write the equation for the production of poly(tetrafluoroethene) from tetrafluoroethene. Clearly show the repeating unit of the polymer.

Answer

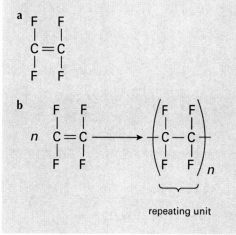

repeating unit

Questions

1 What is the major product for the reaction between propene and hydrogen bromide?

2 Give the equation for the formation of ethanol from ethene.

3 Suggest a use for poly(propene).

2.17 Alcohols 1

OBJECTIVES

By the end of this section you should be able to:

○ recall that alcohol can be made industrially by fermentation

○ describe the conditions used in fermentation

○ understand the advantages and disadvantages of making alcohol by fermentation and by the hydration of ethene

○ explain the term 'biofuels'

○ understand that ethanol produced by fermentation may be a carbon-neutral fuel

○ recall that alcohols can be classified as primary, secondary, or tertiary

Fermentation

- Ethanol is a type of **alcohol**.

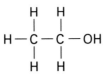

- Alcohols have the general formula $C_nH_{2n+1}OH$.
- Alcohols contain the OH or **hydroxyl** group.
- Alcohols that have more than two carbon atoms can exist as positional isomers, for example propan-1-ol and propan-2-ol. The number indicates the position of the hydroxyl group.

propan-1-ol   propan-2-ol

Worked example

Draw the structural formula of pentan-2-ol.

Answer

pentan-2-ol

Industrial alcohol

- Fermentation can also be used to make ethanol for use as fuels or for industrial purposes.
- The sugar comes from a wide range of sources including sugar beet or sugar cane.
- The reaction is carried out at around 36°C and a pH of 4.5 which helps to prevent the growth of bacteria.

Biofuels

- Biofuels such as wood are fuels that are made from living things.
- Ethanol produced by the fermentation of sugar is a **biofuel**.
- A **carbon-neutral** fuel releases no net carbon emissions to the atmosphere over a year.

Ethanol produced by the **hydration** of ethene is not carbon-neutral because the ethene comes from crude oil and when the ethanol is burnt the carbon is released into the atmosphere as carbon dioxide.

Ethanol produced by the fermentation of corn is also not carbon-neutral. Farmers use large amounts of fossil fuels

- to make the fertilizers used to increases the yields of corn
- to fuel the tractors to plant and harvest the corn

Some carbon dioxide is absorbed from the atmosphere as the corn grows but this carbon is released into the atmosphere when the ethanol is burnt.

However, ethanol produced from sugar cane can be carbon-neutral.

As the sugar cane grows it absorbs carbon dioxide from the atmosphere. The sugar cane is then converted into ethanol. Any waste sugar cane can be used to fuel the process, for example as a fuel for vehicles. When the alcohol is burnt the carbon dioxide is released back into the atmosphere. This could be very important in the future.

Ethanol can be made industrially by the hydration of ethene

$$C_2H_4 + H_2O \rightarrow C_2H_5OH$$

This is an addition reaction. Addition reactions have an **atom economy** of 100%. This is good for sustainable development. All the reactant atoms are made into useful products.

Ethanol can also be made industrially by the fermentation of glucose from plants.

Glucose → ethanol + carbon dioxide

$$C_6H_{12}O_6 \rightarrow 2CH_3CH_2OH + 2CO_2$$

This is a decomposition reaction. Carbon dioxide is also produced by the reaction although this can be used in the manufacture of fizzy drinks.

Wine and beer

- Fermentation on a large scale is used to produce wine and beer.
- The sugar in barley is used to make beer.
- The sugar in grape juice is used to make wine.

Fermentation versus hydration of ethene

fermentation	hydration of ethene
slower rate of reaction.	faster rate of reaction.
yield of ethanol of around 15%.	yield of ethanol of around 95%.
batch process. It is more labour intensive, so labour costs are higher but set up costs are lower.	**continuous process**, so labour costs are less but set-up costs are higher.
atom economy of 51.1%.	an addition reaction with an atom economy of 100%
uses lower temperatures and pressures so uses lower amounts of energy.	uses higher temperatures and pressures so uses higher amounts of energy.
renewable	non-renewable
Distillation is used to increase the concentration of alcohol. The alcohol that is produced has a lower purity so more steps are required for purification.	Distillation is used to increase the concentration of alcohol. The alcohol produced is already purer so purification is easier.

Worked example

Ethanol can be made by fermentation or by the hydration of ethene.

The ethanol produced is then purified.

a How is the ethanol purified?

b Why is the purification of ethanol made by the hydration of ethene easier than the purification of ethanol made by fermentation?

Answer

a Distillation

b Ethanol made by the hydration of ethene is already purer.

Types of alcohol

We can classify alcohols as primary, secondary, or tertiary depending on the number carbon atoms attached to the carbon atom bonded to the hydroxyl, OH, group.

Tertiary alcohols

In tertiary alcohols the carbon atom attached to the hydroxyl group is attached to three other carbon atoms.

A tertiary alcohol

Secondary alcohols

In secondary alcohols the carbon atom attached to the hydroxyl group is attached to two other carbon atoms.

A secondary alcohol

Primary alcohols

In primary alcohols the carbon atom attached to the hydroxyl group is attached to one other carbon atom.

A primary alcohol

Worked example

Draw the structural formula of the tertiary alcohol with the formula $C_4H_{10}O$.

Answer

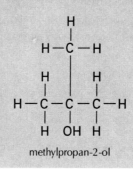

methylpropan-2-ol

Questions

1 Give two ways that ethanol can be made.
2 What is a carbon-neutral fuel?
3 What type of alcohol is propan-2-ol?

2.18 Alcohols 2

Tertiary alcohols

Tertiary alcohols such as methylpropan-2-ol have a hydroxyl group attached to a carbon atom that is attached to three other carbon atoms.

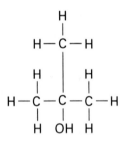

- Tertiary alcohols are not oxidized by oxidizing agents such as aqueous acidified potassium dichromate(VI), $K_2Cr_2O_7$.

  ⓟ As a result if acidified potassium dichromate(VI) solution is added to a tertiary alcohol no oxidation reaction takes place so the orange solution stays orange.

Primary and secondary alcohols

Primary and secondary alcohols are oxidized by oxidizing agents such as aqueous acidified potassium dichromate(VI), $K_2Cr_2O_7$.

Primary alcohols

Primary alcohols are first oxidized to **aldehydes**.

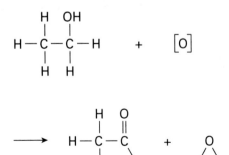

$$CH_3CH_2OH + [O] \rightarrow CH_3CHO + H_2O$$

Notice how the oxidizing agent is written as [O] but it is important that the equation still balances.

As the primary alcohols is oxidized the chromium(VI) ion is reduce to chromium(III) and the solution changes colour from orange to green.

If the oxidizing agents are in excess then the aldehyde is oxidized to a **carboxylic acid**.

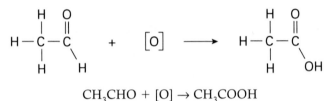

$$CH_3CHO + [O] \rightarrow CH_3COOH$$

As the aldehyde is oxidized the chromium(VI) ion is reduced to chromium(III) and the solution changes colour from orange to green.

Secondary alcohols

Secondary alcohols are oxidized to **ketones** but they cannot be oxidized any further.

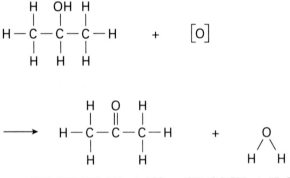

$$CH_3CH(OH)CH_3 + [O] \rightarrow CH_3COCH_3 + H_2O$$

Distinguishing between aldehydes and ketones

Aldehydes and ketones both contain carbonyl, C=O, groups.

Aldehydes

In aldehydes the carbonyl is at the end of the carbon chain.

Ketones

In ketones the carbonyl is not at the end of the carbon chain.

$$R-\overset{\overset{\textstyle O}{\|}}{C}-R'$$

Tollens' reagent

Tollens' reagent, ammoniacal silver nitrate, can be used to distinguish aldehydes from ketones. When Tollens' reagent is added to aldehydes they are oxidized to carboxylic acids and the silver Ag^+ ions in the Tollens' reagent are reduced to silver atoms, Ag, and a silver mirror forms.

Ketones cannot be oxidized so when Tollens' reagent is added to ketones there is no reaction.

Fehling's solution

Fehling's solution is a blue solution that contains a complex of Cu^{2+} ions.

When Fehling's solution is heated with an aldehyde the aldehyde is oxidized to a carboxylic acid and the Cu^{2+} ions are reduced to Cu^+ ions. A brick-red precipitate of Cu_2O forms.

Ketones cannot be oxidized, so no reaction happens and the blue Fehling's solution does not change colour.

Producing polymers from alcohols

Dehydrating agents such as concentrated sulfuric acid or phosphoric acid can be used to remove H_2O from alcohols to form alkenes.

Example

$$\text{propan-1-ol} \rightarrow \text{propene} + \text{water}$$
$$CH_3CH_2CH_2OH \rightarrow CH_3CHCH_2 + H_2O$$

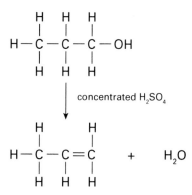

Notice that as water is lost from the alcohol during the reaction this may also be classified as an **elimination** reaction.

Alkenes made in this way could be used as raw materials to make **polymers**.

Here the **monomer** propene is used to make the polymer poly(propene).

Polymers produced from alcohols made by the fermentation of plant materials provide an alternative to using monomers derived from crude oil, which is a non-renewable resource.

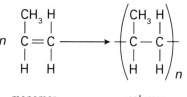

monomer   polymer

Worked example

Three carbonyl compounds X, Y, and Z all have the molecular formula C_4H_8O.

Compound X does not react with Tollens' reagent or with Fehling's solution.

Compound Y is made by the oxidation of methylpropan-1-ol. It reacts with Tollens' reagent to give a silver mirror and gives a red precipitate with Fehling's solution.

Compound Z reacts with Tollens' reagent to give a silver mirror and gives a red precipitate with Fehling's solution.

Suggest structural formulae and names for compounds X, Y, and Z.

Answer

X is a carbonyl that cannot be oxidized so it is a ketone.

It has the structural formula

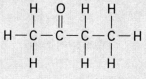

butanone

Y is an aldehyde made from the oxidation of methylpropan-1-ol.

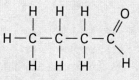

butanal

Y is also an aldehyde.

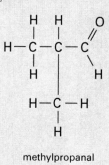

methylpropanal

Questions

1. What is formed when a secondary alcohol is oxidized?

2. Give two reagents that could be used to differentiate between an aldehyde and a ketone.

3. Suggest a dehydrating agent that could be used to form an alkene from an alcohol.

2.19 Analytical techniques

OBJECTIVES

By the end of this section you should be able to:

○ *describe how mass spectroscopy can be used to determine the molecular formula of a compound from the mass of the molecular ion*

○ *recall that certain groups in an organic molecule absorb infrared radiation at characteristic frequencies*

○ *recall that 'fingerprinting' can be used to identify a molecule by comparison of spectra*

○ *use spectra to identify functional groups and identify impurities*

○ *understand how the absorption of infrared radiation by bonds in carbon dioxide, methane, and water vapour leads to global warming*

Modern instrumental methods

The **mass spectrometer** and **infrared spectrometer** can be used to identify compounds.

Compared with traditional laboratory techniques, these modern methods of analysis are:

- faster
- more accurate
- more sensitive
- able to use smaller samples

These spectrometers can be connected directly to computers which can process enormous amounts of information at very high speed. Modern instrumental methods can also identify how much of the compound is present.

Mass spectroscopy

This diagram shows what happens to a sample in a mass spectrometer.

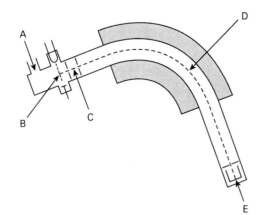

A mass spectrometer

At A the sample is **vaporized**.
At B the sample is **ionized**.
At C the ions are **accelerated** by an electric field.
At D the ions are **deflected** by a magnetic field.
At E the ions are **detected** to produce a mass spectrum.

High-resolution mass spectroscopy

We use high-resolution mass spectroscopy to measure the mass of the molecular ion, M^+, very **precisely**.

Because of the very high precision used we can determine the molecular formula of a compound.

Example

A sample has a relative atomic mass of 30.0688 but is it ethane or methanol?

element	relative atomic mass
H	1.0079
C	12.0107
O	15.9994

Ethane, C_2H_6 has a relative atomic mass of

$$2 \times 12.0107 + 6 \times 1.0079 = 30.0688$$

While methanal, CH_2O

$$1 \times 12.0107 + 2 \times 1.0079 + 1 \times 15.9994 = 30.0259$$

So by using high-resolution mass spectroscopy we can deduce that the sample must be ethane, C_2H_6.

Infrared spectroscopy

Infrared spectroscopy is used to identify organic molecules. The covalent bonds between atoms are like springs. In infrared spectroscopy infrared radiation is used to make the bonds vibrate.

Different types of bonds absorb infrared radiation of slightly different wavelength. By seeing which wavelengths have been absorbed we can identify **functional groups** in organic molecules.

typical wavenumber (cm⁻¹)	bond
750–1100	C–C
1000–1300	C–O
1620–1680	C=C
1650–1750	C=O
2500–3500	O–H in carboxylic acids
2850–3300	C–H
3230–3550	O–H bond in alcohols
3300–3500	N–H

We measure the infrared radiation absorbed using wavenumbers. The wavenumber is the number of waves in 1 cm. The higher the frequency the more waves in 1 cm so the higher the wavenumber.

86

Infrared spectrum of ethanol.

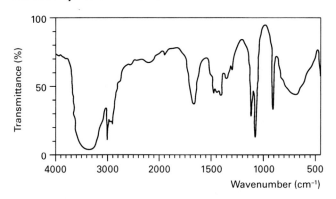

Notice that the infrared spectrum of ethanol shows a broad peak at 3400 cm^{-1} caused by the O–H bond and another at 1100 cm^{-1} caused by the C–O bond.

Useful ideas

Alkanes

Alkanes will contain C–H bonds so they will have an absorption band between 2850 and 3300 cm^{-1} and probably also C–C bonds so they will also have an absorption band between 750 and 1100 cm^{-1}

Alkenes

Alkenes will also contain C–H bonds and will also probably contain C–C bonds. However, they will also contain C=C bonds so they will have an absorption band between 1620 and 1680 cm^{-1}.

Carbonyls

Aldehydes and ketones both contain C=O bonds so they will have an absorption band between 1650 and 1750 cm^{-1}.

Carboxylic acids

Carboxylic acids also contain C=O bonds so they will have an absorption band between 1650 and 1750 cm^{-1}. In addition they contain an O–H bond so they will have an absorption band between 2500 and 3500 cm^{-1}.

Alcohols

Alcohols contain a C–O bond so they will have an absorption band between 1000 and 1300 cm^{-1}

They will also have a O–H bond so they will have a broad absorption band between 3230 and 3550 cm^{-1}

Ethers

Ethers contain C–O bonds so they will have an absorption band between 1000 and 1300 cm^{-1}.

Questions

1 What is special about high-resolution mass spectroscopy?

2 Name three greenhouse gases.

3 How can you use infrared spectroscopy to show that a sample contains impurities?

Worked example

Here is the infrared spectrum of an organic compound.

Explain how this spectrum proves that the compound is butan-2-ol not butanone.

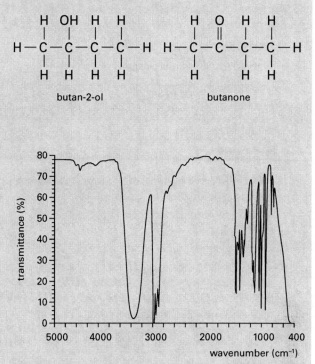

Answer

Butanone contains a C=O bond so we would expect to see an absorption band between 1650 and 1750 cm^{-1} which is absent.

Butan-2-ol has a C–O bond so it should have an absorption band between 1000 and 1300 cm^{-1} which is present. It also has an O–H bond so it should have a broad absorption band between 3230 and 3550 cm^{-1} which is also present.

Impurities

Any impurities in a sample will produce absorption bands that should not be there.

The fingerprint region

The area of an infrared spectrum between 400 cm^{-1} and 1500 cm^{-1} is known as the **fingerprint region** because it is unique for every compound.

An unknown sample can be identified by comparing its infrared spectrum with a database of known infrared spectra to find a match.

HSW: Global warming

Carbon dioxide, methane, and water vapour are greenhouse gases. The bonds in these molecules absorb infrared radiation from the Earth's surface. The radiation is then emitted in all directions. Although some of this radiation is lost into space some increases the temperature of the atmosphere, resulting in global warming.

Further practice

Your examination will involve a combination of structured questions and more open ended longer answer questions. It is essential that you identify everything that you must include in the longer answers. Here is a selection of examination-style questions for Units 1 and 2.

Unit 1 questions

1 a Complete the following table

	relative mass	relative charge	location in an atom
neutron			
electron			

b Write down the number of protons, neutrons, and electrons in an atom of calcium.

c Write down the electron arrangement of the calcium ion.

d Define the term 'relative atomic mass'.

e The data below were obtained from a mass spectroscopy experiment. Use the data to calculate the relative atomic mass of the element. Give your answer to one decimal place.

m/z	90	91	92	94	96
relative abundance (%)	51.5	11.2	17.1	17.4	2.8

f What is the identity of the element?

2 a Define the term first 'ionization energy'.

b Complete the electron arrangement for the Na^+ ion, $1s^2$ ....

c Identify the block in the table to which sodium belongs.

d Write a balanced equation for the third ionization energy of sodium.

3 a Hydrazine, N_2H_4, can be used as a rocket fuel. It reacts with oxygen to form nitrogen and water:

$$N_2H_4(l) + O_2(g) \rightarrow N_2(g) + 2H_2O(g)$$

  i A 2.00 g sample of hydrazine was burned in oxygen forming nitrogen and steam. Calculate the number of moles of hydrazine burnt.

  ii State the ideal gas equation and use it to calculate the volume of nitrogen gas produced from burning 2.00 g of hydrazine at 298 K and 100 kPa (The gas constant $R = 8.31$ J K^{-1} mol^{-1})

b Another compound, X, can also be used as a rocket propellant. It contains 5.88% hydrogen and 94.12% oxygen by mass.

  i State what is meant by the term empirical formula.

  ii Use the data above to calculate the empirical formula of compound X.

  iii Compound X is found to have a relative molecular mass of 34. Determine the molecular formula of compound X.

4 The equation for the reaction of potassium hydroxide with hydrochloric acid is given below. Calculate the volume, in cm^3, of 0.2 mol dm^{-3} hydrochloric acid required to react completely with 0.3 g of potassium hydroxide.

$$KOH(s) + HCl(aq) \rightarrow KCl(aq) + H_2O(l)$$

5 a Complete the following table to show the types of bonding:

	substance	type of structure	type of bonding
i	sodium fluoride	giant	
ii	potassium		
iii	diamond		covalent
iv	bromine		

b Sketch a small section of the diamond structure.

c Explain why diamond is a solid at room temperature and bromine is a liquid.

d Molten sodium fluoride and solid potassium metal are both able to conduct electricity. Explain why this is possible.

6 a The table shows the electronegativity values of some elements

	carbon	hydrogen	fluorine	chlorine	bromine	iodine
electro-negativity	2.5	2.1	4.0	3.0	2.8	2.5

  i Define the term electronegativity.

  ii Use the electronegativity data to label any dipoles in the bonds below:

  H–F   C–Cl   Cl–I

b Draw a dot and cross diagram for CH_3Cl.

c Draw a diagram showing the shape of CH_3Cl and label the bond angles.

d Explain why CH_3Cl is described as a polar molecule while CCl_4 is not.

7 Ammonia will react with boron trifluoride forming $NH_3 \cdot BF_3$.

a Draw diagrams to show the shape of the NH_3 molecule and the BF_3 molecule. In each case name the shape and label the bond angles.

b In terms of electrons, explain how the bond between the BF_3 and the NH_3 molecule is formed. Name the type of bond formed in this reaction.

8 a State and explain the trend in atomic radius of the elements Na to Cl in period 3.

b Examine the graph showing the boiling points of the period 3 elements. Explain in terms of their structure why sulfur has a higher boiling point than phosphorus and chlorine.

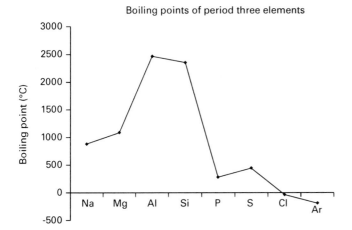
Boiling points of period three elements

9 Mass spectrometry is used to find out about the relative atomic mass of substances.

a Copy and complete the table to show the names of the key steps in the mass spectrometry experiment:

i	an electron gun is fired at the sample
ii	the sample is passed through an electric field
iii	the sample is passed through a magnetic field

b Define the term 'isotope'.

c State the numbers of protons, neutrons, and electrons in the two isotopes of chlorine, ^{35}Cl and ^{37}Cl.

d A sample of chlorine gas was passed through a mass spectrometer showing that the relative abundances of ^{35}Cl and ^{37}Cl are 75% and 25% respectively. Calculate the relative atomic mass of chlorine.

10 a Describe the bonding in, and the structure of, sodium chloride and graphite. Draw diagrams to illustrate your answer. Explain how both of these substances are able to conduct electricity.

b In the late 1800s, research was carried out into the forces that enabled argon to liquefy. State the name of these forces and explain how they arise.

11 Octane (C_8H_{18}) is a hydrocarbon which is obtained from petroleum. It is used as a major component of petrol.

a Name the process used to separate octane from petroleum.

b Write an equation for the complete combustion of octane.

c Octane is mixed with a number of branched chain hydrocarbons in order to make a fuel which burns more efficiently. Draw two possible branched chain alkanes which have the same molecular formula as octane.

d Octane can be cracked in the presence of a catalyst to form hexane, C_6H_{14}, and an alkene. Write a balanced equation for this cracking reaction.

e State the name of the catalyst used in this reaction.

f Hexane reacts with chlorine gas in the presence of ultraviolet light to form 2-chlorohexane:

$C_6H_{14} + Cl_2 \rightarrow C_6H_{13}Cl + HCl$.

i Calculate the % atom economy for the formation of $C_6H_{13}Cl$ in this reaction.

ii Calculate the maximum possible mass of $C_6H_{13}Cl$ that can be formed from the complete reaction of 1 g of hexane with chlorine gas.

I still need to:

Done

- -- ☐
- -- ☐
- -- ☐
- -- ☐
- -- ☐
- -- ☐
- -- ☐
- -- ☐

Unit 2 questions

1 **a** State Hess's Law

The table below shows some standard enthalpy change of combustion values.

substance	$\Delta H_c^{\ominus}$ (kJ mol^{-1})
$H_2(g)$	−285.8
C (graphite)	−393.5
$C_2H_6(g)$	−1560

b Define the term 'standard enthalpy of formation'.

c Draw a Hess's Law cycle for the formation of ethane, C_2H_6.

d Use the standard enthalpy of combustion values to calculate the standard enthalpy of formation of ethane.

e State the significance of the sign of the value of answer **d**.

2 Methane reacts with chlorine in a homolytic free radical reaction.

The initiation step is shown below.

$$Cl-Cl \longrightarrow 2Cl\bullet$$

a What are the conditions required?

b What is a free radical?

The propagation steps for the reaction are shown below.

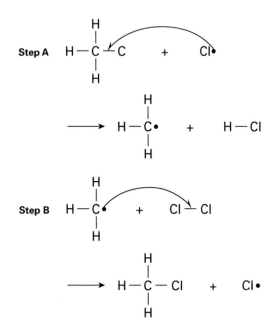

c Give the equation for the termination step that produces a product that does not contain chlorine.

3 The diagram below shows the energies of the particles in a sample of a gas at a given temperature.

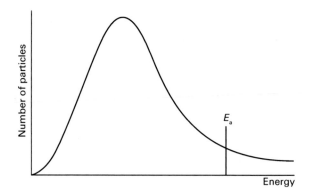

a The activation energy is marked in the diagram. What is the activation energy of a reaction?

b Sketch onto the graph the distribution curve for the sample of gas at a higher temperature.

c Explain why increasing the temperature increases the rate of reaction.

d How does adding a catalyst affect the energy distribution curve of a gas sample?

e Explain why adding a catalyst increases the rate of a chemical reaction.

4 **a** Complete the table below to show that the state at room temperature of the halogens.

element	state at room temperature
fluorine	
chlorine	
bromine	
iodine	

b Explain the trend in boiling point down the halogen group.

c Describe how you could use silver nitrate solution and ammonia solution to distinguish between a solution of sodium bromide and a solution of sodium iodide.

5 The equation below shows how chlorine reacts with water.

$$Cl_2(aq) + H_2O(aq) \rightleftharpoons HClO(aq) + HCl(aq)$$

a State the oxidation number of chlorine in

i Cl_2

ii HClO

iii HCl

b Define the term 'disproportionation' and explain why the reaction between chlorine and water can be described as disproportionation.

6 Two carbonyl compounds A and B have the formula C_3H_6O.

   a When a sample of compound A is warmed with Fehling's solution a red precipitate of Cu_2O is formed.

   Name and draw the structural formula of compound A.

   b When a sample of compound B is warmed with Fehling's solution there is no reaction.

   Name and draw the structural formula of compound B.

   c Compound A is made by the oxidation of compound X.

   Name and draw the structural formula of compound X.

   d Compound A can be oxidized to form compound Y.

   Name and draw the structural formula of compound Y.

7 Propene has the formula C_3H_6.

   Propene reacts with hydrogen bromide to form a haloalkane.

   a State the type of reaction that takes place between propene and hydrogen bromide.

   b Write the name and structural formula of the major product of the reaction between propene and hydrogen bromide.

   c Draw the mechanisms for the reaction between propene and hydrogen bromide.

8 **a** State the trend in solubility of the sulfates of group 2 metals.

   b What would you see if a solution of barium chloride was added to a solution of copper sulfate?

   c Write a balanced symbol equation for the reaction between barium chloride solution and copper sulfate solution, including state symbols.

9 1-bromopropane can react with an ethanolic solution of sodium hydroxide to form propene.

   a State the name of the mechanism for this reaction.

   b What is the role of the sodium hydroxide?

   1-bromopropane can also react with an aqueous solution of sodium hydroxide to form a different product.

   c Name and draw the structural formula of this product.

   d State the name of the mechanism for this reaction.

10 Haloalkanes can be classified as primary, secondary, or tertiary. Draw the structural formula of a primary, secondary, and tertiary haloalkane with the molecular formula C_4H_9Cl.

11 This questions concerns pent-2-ene, C_5H_{10}.

   Pent-2-ene exists as two diastereoisomers.

   a Draw the two diastereoisomers of pent-2-ene.

   b State why this type of isomerism exists.

   c Explain why pent-1-ene does not exist as diastereoisomers.

   d Name the major product of the reaction between pent-1-ene and hydrogen bromide.

I still need to:

Done

- -- ☐
- -- ☐
- -- ☐
- -- ☐
- -- ☐
- -- ☐
- -- ☐
- -- ☐
- -- ☐
- -- ☐

Answers (numerical)

Answers are provided for questions requiring numerical answers only. Molecular models structures can be checked though sites such as Google 'images'.

Further practice questions Unit 1

1 e relative atomic mass

$$= \frac{(90 \times 51.5) + (91 \times 11.2) + (92 \times 17.1) + (94 \times 17.4) + (96 \times 2.8)}{100}$$

$$= 91.3$$

3 a i 0.0625 moles

ii $pV = nRT$

$V = nRT/p$

n (mol N_2) = 0.0625 $p = 100 \times 10^3$ Pa

$$V = \frac{(0.0625 \times 8.31 \times 298)}{(100 \times 10^3)} = 1.55 \times 10^{-3} \text{ m}^3$$

b i

element	hydrogen	oxygen
% by mass	5.88	94.12
÷ A_r	5.88	5.88
ratio	1	1

Empirical formula Ho

ii Empirical mass = 17. Molecular formula is twice the empirical formula HO, i.e. H_2O_2.

4 moles KOH = 0.3/(39.1 + 16 + 1) = 5.34×10^{-3}

moles HCl = moles KOH

volume of HCl = $5.34 \times 10^{-3}/0.2 = 0.0267$ dm³

= 26.7 cm³

9 d Relative atomic mass = $\frac{(37 \times 25) + (35 \times 75)}{100}$

= 35.5

11 f i % atom economy = (120.5/157) × 100 = 76.8%

ii mol C_6H_{14} = 1/86 = 0.0116

max possible moles of $C_6H_{13}Cl$ = 0.0116

max possible mass of $C_6H_{13}Cl$ = 0.0116 × 120.5 = 1.4 g

Further practice questions Unit 2

1 c

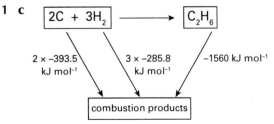

d Enthalpy change of formation = 2 × −393.5 kJ mol⁻¹ + 3 × −285.8 kJ mol⁻¹ − (−1560 kJ mol⁻¹) = −84.4 kJ mol⁻¹

e The sign is negative, which shows that the reaction is exothermic.

5 a i 0   **ii** +1   **iii** −1

Answers to spread questions

Spread 1.01

1 $A_r = \frac{(20 \times 90.9) + (21 \times 0.3) + (22 \times 8.8)}{100} = 20.2$

The element is neon.

Spread 1.03

1 a $n = \text{mass}/M_r = 4/56.1 = 0.07$

b mass = $n \times M_r = 0.15 \times 39.1 = 5.87$ g

c mass = $n \times M_r = 0.32 \times 23.9 = 7.65$ g

d $n = \text{mass}/M_r = 6/40 = 0.15$

   $c = n/v = 0.15/1 = 0.15 \text{ mol dm}^{-3}$

e $n = \text{mass}/M_r = 3.2/36.5 = 0.088$

   $c = n/v = 0.088/(300/1000) = 0.29 \text{ mol dm}^{-3}$

f $n = c \times v = 0.1 \times (25/1000) = 2.5 \times 10^{-3}$

g $n = c \times v = 0.2 \times (21.7/1000) = 4.3 \times 10^{-3}$

h $v = n/c = 0.25/2 = 0.125 \text{ dm}^3$

2 $NaOH + HNO_3 \rightarrow NaNO_3 + H_2O$

   25 cm³   27.60 cm³

   ?   0.15 mol dm⁻³

   $nHNO_3 = c \times v = 0.15 \times (27.60/1000)$

   = 4.14×10^{-3}

   1 mol NaOH reacts with 1 mol HNO_3

   So n NaOH = 4.14×10^{-3}

   $c = n/v = 4.14 \times 10^{-3}/(25/1000) = 0.17 \text{ mol dm}^{-3}$

Spread 1.04

1 Vol H_2(g) = 90 cm³

   Ratio $H_2 : NH_3$ = 3:2

   Vol of NH_3(g) formed = 90 × (3/2) = 135 cm³

2 $V = \frac{nRT}{p} = \frac{2.31 \times 8.31 \times (32 + 273)}{2 \times 10^5} = 0.029 \text{ m}^3$

3 $n(\text{Mg}) = \text{mass}/M_r = \frac{0.2}{24.3} = 8.23 \times 10^{-3}$

   ratio $\text{Mg} : H_2$ = 1:1

   $n(H_2) = 8.23 \times 10^{-3}$

   $V = \frac{nRT}{p} = \frac{(8.23 \times 10^{-3}) \times 8.31 \times 298}{1 \times 105} = 2.04 \times 10^{-4} \text{m}^3$

Spread 1.05

1 % yield = (actual yield/theoretical yield) × 100

   n ethene reacting = 2/28 = 0.07 moles

   theoretical moles of bromoethane = 0.07

   theoretical yield of bromoethane

   = 0.07 × 108.9 = 7.63 g

   % yield = (5.8/7.63) × 100 = 76.0%

2 atom economy = $\frac{\text{relative mass of desired product}}{\text{total relative mass of reactants}} \times 100$

   $= \frac{35.5 \times 2}{2(23 + 35.5) + 2(2 + 16) \times 100} = \frac{71}{153} \times 100 = 46.6\%$

The periodic table of the elements

Key

relative atomic mass
atomic number
name
atomic (proton) number

Example:

1.0
H
hydrogen
1

(1)	(2)												(3)	(4)	(5)	(6)	(7)	0
1	**2**												**3**	**4**	**5**	**6**	**7**	**0**
													(13)	(14)	(15)	(16)	(17)	(18)

(1)	(2)	(3)	(4)	(5)	(6)	(7)	(8)	(9)	(10)	(11)	(12)	(13)	(14)	(15)	(16)	(17)	(18)/0
1.0 **H** hydrogen 1																	4.0 **He** helium 2
6.9 **Li** lithium 3	9.0 **Be** beryllium 4											10.8 **B** boron 5	12.0 **C** carbon 6	14.0 **N** nitrogen 7	16.0 **O** oxygen 8	19.0 **F** fluorine 9	20.2 **Ne** neon 10
23.0 **Na** sodium 11	24.3 **Mg** magnesium 12											27.0 **Al** aluminium 13	28.1 **Si** silicon 14	31.0 **P** phosphorus 15	32.1 **S** sulphur 16	35.5 **Cl** chlorine 17	39.9 **Ar** argon 18
39.1 **K** potassium 19	40.1 **Ca** calcium 20	45.0 **Sc** scandium 21	47.9 **Ti** titanium 22	50.9 **V** vandium 23	52.0 **Cr** chromium 24	54.9 **Mn** manganese 25	55.8 **Fe** iron 26	58.9 **Co** cobalt 27	58.7 **Ni** nickel 28	63.5 **Cu** copper 29	65.4 **Zn** zinc 30	69.7 **Ga** gallium 31	72.6 **Ge** germanium 32	74.9 **As** arsenic 33	79.0 **Se** selenium 34	79.9 **Br** bromine 35	83.8 **Kr** krypton 36
85.5 **Rb** rubidium 37	87.6 **Sr** strontium 38	88.9 **Y** yttrium 39	91.2 **Zr** zirconium 40	92.9 **Nb** niobium 41	95.9 **Mo** molybdenum 42	98.9 **Tc** technetium 43	101.1 **Ru** ruthenium 44	102.9 **Rh** rhodium 45	106.4 **Pd** palladium 46	107.9 **Ag** silver 47	112.4 **Cd** cadmium 48	114.8 **In** indium 49	118.7 **Sn** tin 50	121.8 **Sb** antimony 51	127.6 **Te** tellurium 52	126.9 **I** iodine 53	131.3 **Xe** xenon 54
132.9 **Cs** caesium 55	137.3 **Ba** barium 56	138.9 **La *** lanthanum 57	178.5 **Hf** hafnium 72	180.9 **Ta** tantalum 73	183.9 **W** tungsten 74	186.2 **Re** rhenium 75	190.2 **Os** osmium 76	192.2 **Ir** iridium 77	195.1 **Pt** platinum 78	197.0 **Au** gold 79	200.6 **Hg** mercury 80	204.4 **Tl** thallium 81	207.2 **Pb** lead 82	209.0 **Bi** bismuth 83	210.0 **Po** polonium 84	210.0 **At** astatine 85	222.0 **Rn** radon 86
[223.0] **Fr** francium 87	[226.0] **Ra** radium 88	[227] **Ac †** actinium 89	[261] **Rf** rutherfordium 104	[262] **Db** dubnium 105	[266] **Sg** seaborgium 106	[264] **Bh** bohrium 107	[277] **Hs** hassium 108	[268] **Mt** meitnerium 109	[271] **Ds** darmstadium 110	[272] **Rg** roentgenium 111							

Elements with atomic numbers 112-116 have been reported but not fully authenticad

* 58 – 71 Lanthanides

140.1 **Ce** cerium 58	140.9 **Pr** praseodymium 59	144.2 **Nd** neodymium 60	144.9 **Pm** promethium 61	150.4 **Sm** samarium 62	152.0 **Eu** europium 63	157.3 **Gd** gadolinium 64	158.9 **Tb** terbium 65	162.5 **Dy** dysprosium 66	164.9 **Ho** holmium 67	167.3 **Er** erbium 68	168.9 **Tm** thulium 69	173.0 **Yb** ytterbium 70	175.0 **Lu** lutetium 71

† 90 – 103 Actinides

232.0 **Th** thorium 90	231.0 **Pa** protactinium 91	238.0 **U** uranium 92	237.0 **Np** neptunium 93	239.1 **Pu** plutonium 94	243.1 **Am** americium 95	247.1 **Cm** curium 96	247.0 **Bk** berkelium 97	252.1 **Cf** californium 98	[252] **Es** einsteinium 99	[257] **Fm** fermium 100	[258] **Md** mendelevium 101	[259] **No** nobelium 102	[260] **Lr** lawrencium 103

INDEX